Endorsements

"Did you ever wonder where the spirit and voice went of Abraham Maslow, R. D. Laing, Fritz Perls, Jacob Moreno, Sheldon Kopp, Ira Progoff, Carl Whitaker, Eric Berne, and other therapists who inspired your career? It is alive and well—Brad Keeney is on fire with their spirit! If you missed it then, read it now! *The Creative Therapist* will awaken your memory of why you wanted to be a therapist, why it matters, and how to be a catalyst for change."

Stephen Lankton

Author, Therapist, Editor-in-Chief of the American Journal of Clinical Hypnosis

"This is one of the best books I have read in years about psychotherapy. Brad Keeney re-awakens and extends further the creative roots of American psychotherapy, where a living conversational field is valued above fixed theories and formulas. The writing is delightful, challenging, cajoling, and filled with gems of wisdom—just like a good therapy session!"

Stephen Gilligan, Ph.D.

Author, The Courage to Love *and* The Legacy of Erickson

"This book is a breath of fresh air that reminds us of what therapy is really about, of the healing power of relationships and of the beauty of metaphors and symbols. The stories are unforgettable. A great read for therapists and for all those interested in improving their life."

Cloe Madanes

Co-Author (with Tony Robbins) of The Ultimate Relationship Program

"In *The Creative Therapist*, Brad Keeney doesn't just awaken you. He will shake you up, expand your therapeutic repertoire, and magnify your healing power. Read this and move from therapist to healer."

Carl Hammerschlag, M.D.

Author, The Dancing Healers, The Theft of the Spirit, and Healing Ceremonies

"Brad Keeney lights a fire under the whole therapeutic community. Flying under the rubric of 'Therapy Can Be Fun,' he challenges our conventional approaches and presents a dazzling array of innovative ideas and techniques. Through the use of a theatrical metaphor, he stages wondrous family dramas filled with surprise and unexpected possibilities. Monsters become friends, family skeletons are danced with, secret love is expressed, and obstructing perceptions are turned upside down. This exciting book is sure to stir readers' imaginations and stimulate them to experiment with these unique ideas in their own clinical practice."

Peggy Papp, MSW

Author, The Invisible Web: Gender Patterns in Family Relationships *and* The Process of Change; *Director of the Depression and Gender Project, Ackerman Institute for the Family*

"In this deep and systemic book, Bradford Keeney provides a wisdom invitation: We are invited to be natural and spontaneous creative beings in our living and embrace the art of living that is simply living in love. This book shows that the only kind of encounter that is able to open up a space for mutual respect and self-respect in collaboration and in autonomy is where there is harmony without fear of mistakes and without attachment to some transcendent value. Here the legitimacy of everything that happens is accepted and the past and future are a manner of being now in the present — in the continuous change that results spontaneously in the dynamic architecture of the world that we inhabit. Keeney's creative therapy shows us that everything that happens in the cosmos happens when it happens, not afterward, not before, that is to say, in zero time in the spontaneity of the encounter ..."

Humberto Maturana R., Ph.D.

Co-Director, Instituto Matriztico, Santiago, Chile, and author (with Francisco Varela), The Tree of Knowledge: The Biological Roots of Human Understanding and Autopoiesis *and* Cognition: The Realization of the Living

The Creative THERAPIST

The Creative THERAPIST

THE ART OF AWAKENING A SESSION

Bradford Keeney

Routledge
Taylor & Francis Group
New York   London

Routledge
Taylor & Francis Group
270 Madison Avenue
New York, NY 10016

Routledge
Taylor & Francis Group
27 Church Road
Hove, East Sussex BN3 2FA

© 2009 by Taylor and Francis Group, LLC
Routledge is an imprint of Taylor & Francis Group, an Informa business

Printed in the United States of America on acid-free paper
10 9 8 7 6 5 4 3 2 1

International Standard Book Number: 978-0-415-99703-4 (Hardback)

Library of Congress Cataloging-in-Publication Data

Keeney, Bradford P.
　The creative therapist : the art of awakening a session / Bradford Keeney.
　　p. cm.
　Includes bibliographical references and index.
　ISBN 978-0-415-99703-4 (hardbound : alk. paper)
　1. Creative ability. 2. Personality and creative ability. I. Title.

BF408.K43 2009
616.89'14--dc22　　　　　　　　　　　　　　　　　　　　　　　　2009001386

Visit the Taylor & Francis Web site at
http://www.taylorandfrancis.com

and the Routledge Web site at
http://www.routledgementalhealth.com

Acknowledgments

I want to begin by honoring my first mentor of therapy, Carl A. Whitaker, M.D., whose creative genius, courage, and relational presence was an inspiration to all who knew him. Carl embodied being therapeutic. He didn't need a theory or a technique. Simply being in the presence of his complexity was transformative.

Although other creative therapists have also influenced me, one figure in the history of psychotherapy is a lighthouse. I am speaking of Milton H. Erickson, M.D., whose biographical tribute I was honored to edit with his daughter, Betty Alice Erickson. Erickson ended the dead-end idea that therapy was primarily about insight and interpretation. He took another turn and offered a bright light for all those navigating therapy as the performance of transformation. Those who fail to see his beacon risk being clinically lost in the stormy debates over whose totalizing theory is best, rather than asking how to utilize the unique resources present in each session.

I'm fortunate to have worked with many exceptional therapists, beginning with Art Mandelbaum, Kay Kent, Marian Ault-Riche, and other colleagues at the Menninger Foundation in Topeka, Kansas. I also am grateful to my former colleagues at the Ackerman Institute of Family Therapy in New York City, where I taught and served as the director of research. I express special thanks to Olga Silverstein, Peggy Papp, Lynn Hoffman, Peggy Penn, Jeffrey Ross, and Stanley Siegel, among others, whose conversations contributed to my understanding of therapeutic theory and praxis.

A special note of appreciation is extended to the prodigious pioneering therapist Salvador Minuchin, who encouraged me throughout my career and was responsible for my teaching at the Philadelphia Child Guidance Clinic. There, I was able to be a close colleague of H. Charles Fishman and have more time with Carl Whitaker, who was a visiting professor at the time. Others who reached out during ordeals of controversy to encourage pushing the creative edge of therapy deserve notes of appreciation as well: Lyman Wynne, Richard Rabkin, David Keith, George Greenberg, Insoo Berg, Stefan Zora, Don Bloch, Ramon Corrales, Douglas Sprenkle, Cheryl Storm, Tony Heath, Phoebe Prosky, Debra Denman, Klaus Deissler, Jurgen Hargens, Kitty LaPierriere, Ansie Johnson, Brian Canfield, Jana Sutton, Van Frusha, Lamar Woodham, George Haye, Tim Dwyer, and Eddie Parish.

I carry special memories of the history of brief therapy and family therapy, whose mention serves to acknowledge the contexts of interaction that sometimes made a difference. These include advising Steve de Shazer to write his first book, for which I wrote the foreword; developing the Project for Human Cybernetics with Jeffrey Ross in New York City, where we launched numerous surprises into the profession; engaging in many discussions about systemic epistemology in Italy with Gianfranco Cecchin and Luigi Boscolo; having early talks with some of the "Ericksonians" before they hit the workshop trails (my unconscious tried to get them lost in a New York City subway in order to evoke a deeper appreciation of trance); being a trickster voice of the systemic epistemology movement; proposing some of the first ideas about and against narrative in Australia; giving the first workshop on family therapy in Paraguay, which they celebrated with a live circus; delivering the keynote address to the First World Family Therapy Congress in Dublin in spite of old guard traditionalists' efforts to have me blocked; being booed on stage in New York City with my colleagues Tullio Maranhão and Stephen Tyler for introducing postmodern criticism to clinicians; being heckled so many times for challenging any and all privileged views at therapy conferences that it caught the interest of the *New York Times,* whose reporter Glenn Collins spent weeks recording my stories; and being picked up by South African police and interrogated after rallying therapists to dance for freedom and hanging out in a township.

I should also mention the time I whispered an entire workshop with Olga Silverstein because I literally lost my voice as I celebrated the entry of a woman elder and new leader to the field; conducted a wild constructivist experiment with my students by spreading a rumor that I was writing a major critique of research, but did not, to later find articles published that repudiated that which I never wrote; gave a keynote address to an overly serious narrative therapy conference while showing cartoons and sucking helium; planted seeds with my former students Ron Chenail and James Morris to help qualitative research become a more legitimate method for studying therapeutic communication; organized an historic mental health conference at the Menninger Foundation with Gregory

Bateson that included a healing ceremonial moment for Bateson from poet Gary Snyder; set up a controversial debate between Heinz von Foerster and Stephen Tyler that announced the end of systemic interpretation as a dominant discourse; and sat on a conference panel with Sophie Freud, who publicly announced that my book *Aesthetics of Change* had been one of the most significant influences on her thinking and practice, more so than her grandfather's, followed by my politely and irreverently denouncing what I had written.

Then there was giving Wendel Ray, my graduate student at the time, some cash to load up his car with a copy machine and head to a Batesonian archive overseen by a tree farmer in rural Kentucky, thereby marking his entry to becoming an important historian and archivist; directing several doctoral programs where the importance of the performing arts rather than social science was radically proclaimed and included my insistence upon discussing novels and fiction rather than professional texts in my classes; building a family therapy doctoral program in a matter of months at Nova University, necessarily hiring all my students as the faculty and getting every social service grant available; and teaching an entire course on group therapy at the University of St. Thomas based on using a canned ham and television episodes of David Letterman.

I also enjoy reminiscing over some strange sessions hosted by our academic clinics, from "therapy in the dark" with intermittent spot lighting to the use of "electronic sound effects" and "drive by" car therapy; the surprise of getting a critique of my ideas published, though it was cited as being authored by my dog; having the lights on a stage in Brazil explode at the moment I was delivering a paradoxical prescription in a family interview; discussing the nature of ecstatic shamanism with Cloe Madanes as we wondered whether someone might overhear our conversation; being warned by leaders of family therapy in Buenos Aires that hanging around traditional healers would ruin my career; years later giving a keynote address on traditional healing to the World Congress on Psychology at Puebla, Mexico; working the streets, churches, and sweat lodges with African Americans and Native Americans, trying to help urban gang members in serious trouble with the law; dropping out entirely from the mainstream of the profession to be immersed in diverse cultural healing traditions, while conducting therapeutic sessions whenever a village crisis called for it; and being rediscovered as alive and well by Jeffrey Kottler and Jon Carlson, who wrote my biography, which led to numerous keynote addresses and my present return to clinical practice and teaching.

I wish to send a big hug of appreciation to Carl Hammerschlag, a leading creative therapist whose brotherly support has made a difference in my moving forward with this work.

Before wrapping this up, I want to thank all the institutes that have hosted my teaching over the years, throughout the United States, Europe, South America, Japan, Australia, and Africa. There are too many places to list. My students also have been important colleagues along the way, and many of them are now teachers and leaders of the field. I want to single out the ongoing support of Frank Thomas, whose contributions to brief therapy always are circumscribed by healing humor.

I have emphasized the therapists who have contributed to my clinical work, but I also need to at least mention my main academic mentors who led me down the roads of how to face complex ideas. These include Gregory Bateson, Heinz von Foerster, and Stephen Tyler. They helped me "learn to learn" and demonstrated that it is both necessary and foolish to challenge the orthodoxies, including one's own.

For over a decade, my experiential world was taken apart and put back together again by the teachings of many remarkable cultural healers from the rainforests of the Amazon to Asian cities, South American villages, Balinese compounds, and remote African habitats. They taught me that an "awakened heart" is necessary for therapy to be authentically transformative. I am particularly indebted to Ikuko Osumi, Sensei, one of the most revered healers of Japan. She adopted me like a son, taught me her maternal ways, and passed on an ancient Samurai method of healing. She is one of the most remarkable human beings I have ever known. To the Mothers of the St. Vincent Shakers, part of an old Caribbean spiritual culture, who placed me in their ways of rebirth and dedication to helping people, I shout out my devotion. Many of my teachers over the last couple of decades were women elders from cultures where there are neither theories nor libraries. They taught me that relationship and mothering, rather than jousted interpretations, constitute the soul of therapy. Most specially, I bow before the women elders of the Ju/'hoansi Bushmen, as I am most grateful for their deep caring over all these years, openly inviting me into their relational way of knowing and being.

This work would not be possible without the support of my present colleagues at the Center for Children and Families in Monroe, Louisiana, where I serve as Clinical Director. Thank you Cindy Murray for your servant leadership and entrepreneurship. Hats off to Curtis Eberts, Matt Thornton, Pearl Wong, Ashton Hines, Greg Guthrie, Kelly Wright, and everyone else on our team. You have provided me with a warm, kind, and generous home for creative therapy. Thanks to Scott Shelby, Adam Mathews, Marc Fager, Sarah Hellen, Ritchie Sheridan, Amanda McMullen, Todd Gunter, Daisy Work, Kristy Walters, Kristi Lawson, Kathy Jackson, Sonji Tarver, Kayla Fanguy, Sabryna Herring, Crystal Goodman, Aaron Beeson, Kevin Shelby, Jonathan Robert and all the other therapists who work with me in Louisiana.

A special thanks to my colleague Professor Alfonso Montuori and the students at the doctoral program in transformative studies at the California Institute of Integral Studies in San Francisco. I am appreciative of the valuable suggestions and steadfast support voiced by Connie Scharff, the case transcriptions contributed by Laura Ehmann, and the numerous improvements to the text recommended by Peter le Breton, Don Arispe, and Frank Strona.

Before closing, I want to express gratitude to Justin Moore and Lewis Moore for reaching out years ago when I had all but disappeared in the hinterlands of remote fieldwork. They invited me to annually visit their university as a sort of visiting artist of therapy, conducting and discussing cases through the week. This book would not have been created without your involvement and nurturance. Of course, it was the daily strawberry shortcakes at the Bulldog Restaurant at Bald Knob, Arkansas, that made everything come together with whipped cream and nuts on top.

Finally, I am most deeply grateful to my wife, Mev Jenson, and our family, who have always stood by my side, whether we were heading out to a jungle, rainforest, ancient city, research institute, university, or rural clinic. Our family life, with all its adventure, tears, laughter, and love, has been an inseparable part of this book.

Contents

Part III   Therapy of Therapy

Preface

I love this book! The very first words in the Introduction captured first my imagination, then my full attention. *The Creative Therapist: The Art of Awakening a Session* provides desperately needed information and significant validations for all psychotherapists, regardless of the theoretical approach they follow or don't follow. Brad Keeney clearly shows in the accompanying DVD, and in the transcriptions of those sessions complemented by some of his thoughts, that therapy can have enormous creativity. It can involve clients in ways they had never before considered, broadening their views of the world and, more importantly, of themselves. In doing this, Keeney inspires us all to have more courage to awaken our own natural abilities and vibrant expression in clinical sessions. This emphasis on creative therapeutic performance and transformational experiences demonstrates the flexible and open-ended direction clinical practice should follow, rather than any overreliance on static theory and rigid rules.

The Creative Therapist is filled with truly entrancing ideas, evocative word play, and fascinating clinical strategies, all based on unique yet practical frameworks. Keeney shows how we can draw upon our own often overlooked resources while simultaneously helping others access resources they may have never considered. This calls forth a more productive, independent life for the client, filled with laughter, happy surprises, and connectedness with others.

Keeney's imaginative and playful clinical sessions are transformational experiences for the clients, as well as for the therapist. His creative clinical work helps liberate us from interpreting what's wrong with people and shift to underscoring what's right and interesting. Then we can enthusiastically and truthfully applaud even their tentative efforts while mobilizing them to grow further. This is the simple and yet complex mission of any creatively awakened session.

Keeney brings a broad and unique background to the creative performance of therapy. He is an academic professor, accomplished musician, anthropologist of cultural healing traditions, social cybernetician, improvisational performer, and licensed therapist and supervisor. The combination serves him well and he hits the bull's eye once again, creatively awakening a profession that too easily forgets its performance-oriented roots and goals.

In the clinical sessions, Keeney shows how we can be more effectively involved *on stage* with clients' ongoing drama while remaining more open to the possibilities that anything is possible and anything can happen. He gives us a front-row seat in the theatre of creative transformation. Expect to be moved by the unexpected and uncommon movements of a client's (and therapist's) natural and emergent creative expression and growth. Here, artistry is brought back to the practice of psychotherapy.

Keeney's therapeutic work beautifully merges a sensible and practical approach with the creative atheoretical strategies that resonate with those of my father, the late Milton H. Erickson, M.D. It is a delight to read, and to see, therapy done in creative ways that so clearly carry forth the essence of Dad's work. Keeney completely understands therapy as a context filled with creative resources that foster retrieval and actualization of dormant potential and expressions of hidden talents. Erickson also knew this and focused on growth and transformation rather than on problems, solutions, or narratives of explanation. Over and over he emphasized that whatever is presented should be fully utilized. Part of the therapeutic process is the development of a favorable setting for the growth of those capacities— exactly as Keeney explains, demonstrates, and invites us to cultivate and activate within ourselves and our sessions.

The Creative Therapist is a radically important call asking us to draw upon utilization, improvisation, and creation, rather than clichéd understandings and protocols, to guide clinical work. This major and important contribution shows us how therapy can now be more fully accepted and practiced as a realistic complex art form. I celebrate the arrival of this vital and vitally alive book and believe it can help each of us access, trust, and incorporate the deepest sources of creative wisdom we each carry within.

Several years ago, Brad and I edited a book, *Milton H. Erickson, M.D.: An American Healer*, honoring the work of my father. It reiterated to us both that therapy is principally about opening your heart so that a connection or link can be made to a complex, unknowable, unspecifiable wisdom that facilitates an awakening of the therapist's and client's creative interaction. Keeney's work is inseparable from my father's most important teaching: "You let the client grow." It adds the far-less-often noticed and rarely verbalized other part of this wisdom: "You also let the therapist grow."

Erickson is legendary for the many ways he practiced both of these teachings. Most of his students were given unusual assignments as a part of their work with him—for example, climb Squaw Peak or go to the Botanical Gardens and find the boojum tree. They would come back, perhaps not consciously certain of exactly what they had accomplished but absolutely sure they had somehow changed for the better. *The Creative Therapist* encourages all of us—clients as well as therapists—to never stop traversing the least known trails of the rarest traveled journeys toward surprise and creative exploration, both in our sessions and outside them.

Creativity with Erickson's clients is also well-known. Most readers of Erickson's work know about the patient who kept his dog at our family home for years. The man built a doghouse, and my 8-year-old son helped him paint a sign for it with the dog's name. The patient had the capacity for dedicated love—a dog rescued from the pound filled the bill exactly, metaphorically and literally. I can only imagine what helping a little boy paint the name sign for the dog's new home meant to the patient. It was clear my son felt a great connection to the man, as well as respect and regard, as they worked together to create something new for a new life situation. The bond they created has lasted for each over the decades.

I am delighted that *The Creative Therapist* celebrates this kind of tender relational bonding and asks us to honor and use it as the bridge that carries us into the respectful depths of connectivity with others.

I also remember Dad's giving me "lessons of life"—which is actually what much therapy is—in ways that resonated deeply. When I was in college, he was teaching a workshop that I attended as a hypnotic demonstration subject. Dad and I were invited to a faculty party. I wanted to go; Dad was tired and didn't think it worth the effort. But I begged, so he took me. When we got home, I said, "You were right." He looked surprised and asked what he was right about.

"The party wasn't much fun," I said. "It was crowded, not very good food, just not much fun." He was astonished. "I had a lot of fun," he said. Then he added with intensity: "Why wouldn't I? If I am going to do something, I might as well enjoy it." That remains inside my being. It's a maxim I tell my own clients.

The Creative Therapist succeeds fully in bringing us further inside and outside the mystery and mastery of therapeutic art. It fosters growth, relationships, and having a lot of fun, because after all, as Dad asked, "Why shouldn't we?" Best of all, it does so with full belief and expectations that all therapists already have the resources and talents to help awaken the productive and unique growth of each of their clients. And we do.

Betty Alice Erickson, M.S.

Introduction

Therapy: Dead or Alive?

> If you cannot get rid of the family skeleton, you may as well dance with it.

George Bernard Shaw

Imagine writing a prescription for a client to take Rice Krispies three times a day, providing an opportunity for her to learn how to listen to some *snap, crackle, and pop!* How about having an adolescent sleep over the drawn image of a light bulb? Or having a family throw a party for a deer mounted on the living room wall? Inventing a "funny medicine" for a child that helps him enjoy his fears and concerns? Having a couple spend a romantic night in their barn with an outdoor fireworks finale? Encouraging a client to make an eyeglass holder for invisible glasses so she can acquire spiritual sight? Suggesting that a high school counselor set up a mini-theatre in her office? Welcome to the creative ways in which clinical sessions can be awakened.

I am not the first to propose that therapy must be creatively alive for it to be engaging and effective. Like theatre and other transformative arts, if a clinical session feels boring or dead, it probably is. Furthermore, if it isn't alive, it isn't going anywhere. I propose that "creativity"—rather than theory, method, technique, or research—is what awakens meaningful and transformative therapy.

Creativity encourages inspired presence rather than stale imitation. It embraces the process of developing something new, uncommon, and unique. Not simply "new for new's sake," but tailor-made both to fit and help liberate the immediate circumstances. Rather than replicating or reproducing a template to be hammered onto every clinical session, creative therapy custom-builds a therapeutic encounter as the occasion calls it

1

forth. Respecting the authenticity of each moment, it brings about original clinical work.

If we believe each human being is unique, along with all the social interactions and contexts that hold the performance of everyday living, it follows that every clinical session should hold the possibility for conceiving a uniquely invented therapy. With this outlook, the creative therapist is ready to create, compose, construct, form, parent, give rise to, grow, bring forth, bring about, and bring into being an authentic, made-in-the-moment, one-of-a-kind session.

At the same time, therapists (and all human beings) also share much in common. What is important is that we feel free to express this shared commonality in our own unique way, tone, flavor, rhythm, coloring, or movement. Whether our therapy appears as a recognizable form or as radically idiosyncratic is less the point. What's more important is whether it is creatively alive, that is, dynamically expresses our unique touch and authentic contribution. Creative therapy challenges both clients and therapists to be more fully alive and courageously responsive to their present interaction. It asks therapists to tap deeply into imagination and have the commitment freely to flow with improvisation. Awakened therapy encourages uncommon creative presence that aims to foster enacted transformation. It welcomes inspired and unexpected thought and action into each session.

How can we get the creative spark ignited and voiced? How can such moments foster therapeutic change? How can we prepare ourselves to be more creative in psychotherapy? How can we give birth to vibrant transformative change in a therapeutic encounter? How do we hold and foster existential growth? And when things aren't working, how can we utilize impasses, inflexibilities, challenges, and symptomatic experience to evoke creative movement?

While recognizing the value and importance of diverse models, schools, and paradigms of psychotherapy, we must learn how to move creatively with our clients or our sessions will go nowhere. We must learn to go beyond the guidance initiated by previous clinical trainings. An emphasis on creativity provides a revitalized reframing of therapeutic conduct that is applicable to all systems of practice. No matter what therapeutic orientation one practices, it must breathe and circulate creativity in order for sessions to come alive. Anna Freud reminds us in her well-rehearsed line: "Creative minds have always been known to survive any kind of bad training."

I am not saying that all schools of therapy (including meta-school and anti-school approaches) are flawed or without value simply because they have become institutions with orthodox understandings and prescribed rules of conduct. They provide a structure from which creative expression can feed. Like the jazz performer who uses a structured sequence of notes called a *melody* as the basis for embellishing, modulating, or deconstructing, a therapist should feel free to play with the melodies the various schools prescribe. Creativity is often at its improvisational best when it paradoxically uses a structured form to play with.

Outside observers may think that a performance was creative while staying within the recognizable boundaries of a therapeutic orientation or that it broke out and went into free form. The same is said about jazz performance. The point here is that I am not calling for us to discard schools of therapy. I am calling for *therapy jazz*—a performance-based practice that, like music, requires technical know-how and familiarity with the rules of music in order to have the chops to make great improvisational music. Creativity doesn't come out of thin air. It is not babbling noise. It always has deep roots in traditions of knowing and expression.

Following my colleague Alfonso Montuori (1997, 2003, 2005), we must also remember that "creativity" is neither a psychological trait nor a sociologically induced phenomenon. We should be careful about speaking of either "creative people" or "creative situations." When we talk that way, we are easily tempted to rhetorically slip into committing the fallacy of misplaced concreteness or inappropriately using the abstraction to signify a noun, thing, or factor localized inside or outside a human being. Creativity points toward a process of creating. It both originates in and emerges from a complex, circularly intertwined choreography that moves everything from the dancers to the audience, stage, and choreography itself.

In this book, creativity will be regarded as a process metaphor for the radical constructivist view that emphasizes invention, construction, improvisation, and creation. From this perspective, transformation and creativity are indistinguishable. When a clinical session is authentically and wisely *construed* to help transform the contextual situation at hand, it is "therapeutically creative." In therapeutic work, an emphasis on creativity is a call for the transformative presence of a therapist's (and client's) unique personal resources and talents, imagination, unpredictability, spontaneity, and improvisation. Creative therapy is always changing and evolving, ready to utilize what both client and therapist bring to the table. Again, the ultimate creative goal is to deliver a unique therapeutic approach for each unique situation. Sometimes this may appear to be novel expression within a particular school of therapy. At other times, it may appear to break out of the constraints of any or all schools and models.

The mental health professions critically need transformative ideas that suggest how to go outside habituated ways of theorizing about therapeutic change in order to bring forth innovative possibilities. One of my mentors, Carl Whitaker, a founder of family therapy, warned: "I have a theory that theories are destructive … all theories are bad except for the beginner's game playing, until he [or she] gets the courage to give up theories and just live, because it has been known for generations that any addiction, any indoctrination, tends to be constrictive and constipating" (1976, pp. 317–318). Neill and Kniskern interpret Whitaker as saying that theory "destroys creativity and intuition and eventually destroys the therapist" (1982, p. 317). Any form of overreliance upon theory and understanding will likely hinder the natural creative abilities that stand ready to be expressed by a therapist.

It is important not to take anti-theory proposals and make them another stuck theory. Even with that warning, it is impossible to be a theory-free therapist. At the very least, as Peter le Breton (personal communication, October 12, 2008) suggests, the theories a therapist met once upon a time are now residing in the "back seat in the substratum of the therapist's mind." The more important difference has to do with where the therapist's theories are standing, sitting, or resting during therapeutic conduct.

For the beginning therapist, theories and models dangle in and out of consciousness as the neonate clinician tries to figure out how to make sense of a clinical situation. As the therapist matures and develops greater skill at making meaning, the knowing should be able to more easily slip away from conscious residence. The problem is that we easily get too dependent on clinging to theory and understanding, that is, keeping them too centered in consciousness, and find ourselves losing access to the spontaneous flow of play and invention. Perhaps most of therapy (and our culture) is addicted to rigidified ways of holding on to understanding and this interferes with and deadens creative performance.

When we each become more of a nontheoretically conscious therapist, this does not mean that we are inappropriately unrestrained and problematically chaotic. We seek transformation, a journey that starts with being lost or stuck and then finds a way out, proceeding to advance toward a more generative circumstance. A collection of clinical sessions and case studies are used to illustrate how creativity can be brought into therapy. The cases span a wide range of therapeutic situations, including working with children, adults, couples, and families. Each study shows the movement that guided the creative unfolding of each session. Finally, several cases that enact a "therapy of therapy" are presented of innovative work with therapists, helping them develop their own unique creative style of therapeutic presence and performance.

I have not forgotten the first clinical session that I ever witnessed. I was a young scholar attending a workshop conducted by Carl Whitaker. He was seeing a family for reasons I no longer recall. What was striking about the case was how he asked a question that had no readily apparent connection to what the family was discussing. In the middle of the family's report about how troublesome their lives had become, Carl bent over and locked in on the father, saying, "I am seeing a refrigerator." Shocked by Whitaker's off-the-wall comment, the man replied, "How did you know that my family is in the refrigeration business?" Whitaker's intuition and voicing of a primary process association brought forth a creative metaphor that not only exposed family history and business but also began a thawing and releasing of more creative expression in the session. It also was the beginning of the family's successful onward movement with their lives.

I later asked Whitaker what he was doing in that session. He said that, at that particular moment, he simply "stopped doing therapy" and instead chose to "be more fully alive, trusting my own irrationality and craziness."

"The key to therapy," he explained, has to do with "breaking out of our own prisons of reason." That spontaneous moment in Whitaker's session not only brought life to the family that came to see him, it sparked my own fascination with how improvised creativity, rather than frozen theory, directs the art of awakening clinical sessions and life in general.

Over the years, Carl Whitaker and I became close colleagues. He encouraged me to become a therapist and an outspoken academic with an emphasis upon articulating the dynamic, interconnected, circular nature of relationship and family systems. At the time, I was interested in the cybernetic thinking of Gregory Bateson, Heinz von Foerster, Humberto Maturana, Francisco Varela, and other scholars. Whitaker and I pledged that we would foster the growth of clients *with* therapists as opposed to the "fix 'em up" mentality prescribed by simple reductionist models for therapeutic conduct.

He once said to me, "There are very few of us seriously fighting the trivial and embarrassingly naïve models of therapy. We grow through an experience of our 'primary process connection,' not by making rational sense of anything. Now join me in this battle." Though I did not fully understand Carl's invitation at the time, perhaps now I am closer to accepting the mission to help free therapy from the tame categorical boxes of psychological, biological, psychiatric, archetypal, sociological, family, narrative, new science, postmodern, cosmological, and other theories that more often rationally imprison rather than experientially liberate the contributions that creative healing encounters can bring to the lives of clients and therapists.

As I stepped onto the performance stages of therapy, I learned more about another master of improvisation who avoided any emphasis upon overly concretized maps, models, or theories. As a young psychiatrist, he had encountered a so-called schizophrenic who claimed to be Jesus of Nazareth. The patient stood all day in a mental hospital making a back-and-forth motion with his arms. The therapist went up to him and said, "I hear you are a carpenter, and it looks like you are missing a saw." He then arranged to have a saw placed in the patient's hands and some lumber placed beneath them. The patient's back-and-forth motions, along with his identity of being a tradesman from Nazareth, now positively fit into the resourceful context of a "carpenter-in-action" as opposed to the impoverishing context that redundantly framed more proof of an "irrational psychotic." The patient continued working with the tools of carpentry and eventually left the hospital and became employed as a cabinetmaker. The therapist in this case was Milton H. Erickson, who was called the "Mozart of therapy" by Gregory Bateson. The client was a lost carpenter in search of the right tools and the right context. (Erickson & Keeney, 2006, pp. 13–14.)

There is another story, quite similar to the above case, that involves Ronald Laing. In this situation, Laing was asked to hold a consultation at a conference of therapists. He was presented on stage with a so-called catatonic patient. After the young woman was introduced, Laing wasted no time in saying, "I hear you have a talent for being still." He proceeded to invite her to get paid for her talent, suggesting that she might make a good model at an art studio. Like Milton Erickson, Ronald Laing moved the context from one that previously held the perceived "abnormal behavior" to a different context where the same behavior no longer appeared as abnormal or deficient but became illumined as a resourceful talent. The problem never was the behavior (or expression), but the context that held it. When Laing made a contextual shift, the woman was able to have a conversation about her life that could help her move forward in a more productive way.

For years, as the director of a couple of family therapy doctoral programs, I taught that what was essential about therapy was contained in these clinical stories. I also added the stories of Olga Silverstein, my former colleague at the Ackerman Institute of Family Therapy. I had written detailed analyses of her work, including the treatment of a family with a daughter who had migraine headaches. By altering how the family organized themselves around worrying, the headaches went away. Silverstein's work showed, along with that of Whitaker, Erickson, and Laing, that the art of therapy has more to do with flipping the contexts that house experience than it does with attacking problems, finding solutions, or editing narrative (mis)understandings.

Creative therapy is a call for the *therapy of therapy*. By this I mean that therapists, like their clients, can get stuck and require emancipation. A frozen clinical practice needs a creative fire and a light that reveals new possibilities for how it can become more alive. Therapy—including both medical and nonmedical models—has become a holding tank that feeds problems as well as their codependent attempted solutions and narrative rationales. I shall argue that therapy must become less the proving ground for theoretical belief, less preoccupied with hermeneutics, less subservient to the hierarchical influence of science, less procedure and model oriented, less schooled, less concerned about political correctness, and less professional in order to be more therapeutic. The more transformative alternative requires a creative dissolution and deconstruction of any context that comfortably holds problems and solutions, including any talk, explanations, or narratives about them. We need to become architects of a new stage that directs its spotlight on the resources, gifts, and transformative lessons delivered by any particular experience, whether it voices suffering or joy. Most importantly, we need to get on this stage and emphasize the *performance* rather than the interpretation of transformation.

Another way of saying this is that the art of awakening a session requires moving out of all contexts that perpetuate the therapy game. With our clients, we must walk into an alternative *theatre of creative transformation*, where we play with, rather than brood over, whatever is presented. Here, all experience is grist for the alchemical processes that utilize, rather than cannibalize, whatever life and death bring to the table.

In the chapters that follow, therapists are asked to minimize theorizing, pay less respect to scientific (and other abstract) renderings of therapeutic process, and still the diagnoses, explanations, and understandings that paralyze spontaneous, existential growth. Instead, a revitalized performance is encouraged that asks for the natural flow of improvisation guided by the nontechnical embodied know-how of how to be authentically alive. The call for creative therapy is nothing less than a call for healing wisdom, the most authentic way of being therapeutically present. In doing so, we should be free to draw upon any tradition that contributes to our being servants of transformation with one another. As such, we may consider all the transformative arts from poetry to dance, music, comedy, theatre, literature, sport, cooking, painting, and sculpting, among all the other names that point to the different ways we express our deepest passions for living.

The separation and isolation of therapy as a so-called "science" (perhaps better named a pseudoscience) disconnected from the other expressive and healing arts has led to a dysfunctional impasse in our ability to evoke creativity in the very occasions that need it most. Let us announce a revolution that emphasizes *performance,* a takeover and makeover of psychotherapy, reframing it as a significant art that is creative, transdisciplinary, and transformative. Let us drop the singular habit of punctuating pathology—whether it addresses genes, biochemistry, neurology, psychological process, fantasized internal dynamics, learning, social interaction, or cultural narrative—and refocus on a resourceful polyphonic diversity of expressed experience.

Among other things, we need to learn to sing and dance with our clients, laugh and weep more often, touch without fear, and feel freer to get more lost in order to be found. It is time for therapy to treat itself. It is in need of healing and transformation. This is not a one-time fix, but a never-ending preparation for creative expression in every single session. Let therapy commit to being a higher order therapy of therapy and a transforming art of transformation. Let us first heal and bring to life our profession before we dare continue trying to heal those who bring suffering and pain.

My own search for creativity in therapy began with my first professional position in the field. I held the invented title "communications analyst." My job was to observe the clinical cases of distinguished therapists at the Menninger Foundation and to articulate the patterns of communication they used to foster change. I eventually moved from Topeka to become the director of research at one of the founding institutes of family therapy in New York City, the Ackerman Institute for Family Therapy. There, I continued studying the communication patterns that organize successful therapeutic transformation. I was fortunate to work with many remarkable creative therapists from leading institutes and universities throughout the world.

During this time, I was a young family therapist, a jazz pianist, and a cybernetic theorist (yes, a theorist). My earliest publications were regarded as classics in the fields of family therapy and brief therapy. The therapists I worked with were the elders and leaders (and often founders) of these fields. To my surprise, I learned that for the most part these innovators of therapy had not developed a formalized articulation of what they were doing in a session. They had to come up with theoretical explanations professionally to account for clinical action, but that came after they found their own unique way of conducting therapy. They naïvely assumed I might be able to help them explain what was going on.

Those who read the books of the great therapy originals are likely to believe that the clinician's theory came before the practice was created because that is the sequence in which they are taught and written about. A seldom told secret is that therapies are usually first invented, brought about by trial and error tinkering, and that theoretical explanation happens afterward. I am speaking about the pioneers, the great contributors to therapeutic practice, not the secondary contributors (and populists) who come along later with published works and bureaucratic structures for institutionalizing the field. This applies to all therapies (from the Greek *therapeia,* meaning curing and healing), from acupuncturists to body workers and allopathic physicians. In the early days of medicine, even allopathic horse and buggy doctors were called "empirics"; they learned what worked through trial and error.

The original inventors or creators of therapeutic approaches often found themselves in a workplace where their clients were not responsive to the available forms of treatment. Whether troubled or incarcerated adolescents, schizophrenics, anorectics, or families of the slums, "impossible to treat" clients gave a therapist an opportunity to try something new. There were essentially no strict rules for treatment because nothing seemed to work. In this situation, empiricism was encouraged—that is, trying one thing after another until the therapist found something that might work. Tinkering enabled a therapist to draw upon his or her own wits and try out different ways of interacting. This experimental attitude constitutes the context for invention in therapy. The original schools of family therapy, in particular, arose out of working this way with various so-called "impossible" client populations.

I cannot overemphasize how important it is to work with difficult clients in order to evolve one's therapeutic approach. Many of the more mundane (though no less important) counseling challenges—from meeting a couple that complains of not being able to communicate, to adolescents who don't do their chores in a timely manner, to someone struggling to make a career decision—do not foster the invention and validation of novel therapeutic orientations. They do not challenge your own assumptions or push you past your comfort zone.

It is easy for therapists (including pop television therapists) to be deluded into thinking that they know what they are doing if they do not face tough problems. Face an impossible case, and overly simplistic and rational counseling must give way to more complex and what often seems irrational therapy. Schizophrenics, anorectics, and delinquent youths are less responsive to placebo interventions and are unlikely to placate clinicians. They require someone who has learned how to face complexity with graceful transformative engagement.

When I directed doctoral programs in therapy, I made sure that pedagogy centered on live casework. Our clinics sought social service grants for marginalized populations and those with difficult problems, from sex offenders to addicts, impossible adolescents, and those involved in violent interactions. With these clients, therapy trainees had a better chance of knowing what was working and what was not. Similarly, in my own experience, I have preferred to work with the most difficult "impossible" cases because they allow you to more easily grow your creativity as a therapist. This is not to say that I only work with the most difficult cases; I work with the whole range of clinical challenges. My point is that we learn best when we are placed in situations that require us to do something different.

As a teacher of therapy, I argued that theory-directed training was a less-productive learning medium than case-directed training. The latter requires that you learn from interacting with clients rather than trying to fit a session inside a theoretical (or procedural) map. When I conduct workshops, I insist that the cases do the teaching. I prefer having a live clinical consultation each morning and afternoon. What clinically takes place directs what and how we talk about creativity in therapy.

This is not a proposal to have no theory in teaching therapy. I am suggesting that we should provide as much time for experiencing a session as the time that is spent with books. Even today, most university clinical training programs devote less time to clinical work than to talking and reading "about" therapy. What we need is to devote equal time to theory *and* practice so as to maintain a healthy and generative interaction between them. To get there, our polemics may require a strategic call that sways our attention to what has been less validated. Perhaps we must say something provocative like "theory hinders clinical work" because it helps shove things back into place so that both knowing and not-knowing can dance together as equal, though contrasting, sides of a relationship.

An emphasis upon live clinical teaching also helps frame therapy as a performing art rather than as a scientifically based method for eradicating disease and deficiency. I believe therapy took a wrong turn when it housed itself inside the social and medical sciences. Imagine how different therapy would be if it resided within the performing arts. It would certainly be more creative, and that difference would make it more alive and transformative. Do not regard the arts as any less serious or potent than the sciences. Art holds our deepest ways of knowing and our most profound forms of expression. Anyone approaching the existential blood and guts of human suffering will be better prepared coming to it with the depth that art provides.

Sometimes our profession speaks of both the art and science of therapy, but it is usually done with science being the dominant discourse and hierarchical method of legitimacy. This hegemony may have nearly destroyed creativity in therapeutic practice. To draw an analogy with another discipline, we can easily agree that there is a science (and engineering) to playing the piano. The instrument must be built and tuned according to the laws of physics and the musician must be alive according to the laws of biology. Even though that is the case, the transformative effects of music have less to due with science and more to do with the creative flights of human interaction, unfettered by the constraints of scientific logic or even reason itself. Music, rather than the production of sound, goes beyond science and rationality. Similarly, therapeutic healing, rather than the production of specific sequences of words and actions, goes beyond common sense.

When I directed a doctoral program in family therapy at Texas Tech University, I made an announcement that the program would not be contextualized by social science but would be primarily considered a performing art. The family social scientists in the department were not pleased with this political move because it potentially threatened their authority in the workplace. But the change in contextualizing practice as a performing art freed both faculty and students to grow their own creative forms of expression.

At that time, I wrote a book entitled *Improvisational Therapy: A Practical Guide for Creative Clinical Strategies* (Keeney, 1990). It emphasized that the performing art of therapy was at its very core an improvisational form of expression. Because we could never predict what clients would utter, we had to abandon a fixed script and be ready to improvise a response that fit the ever-changing and always unpredictable nature of therapeutic conversation.

The foreword to *Improvisational Therapy* was written by Professor Stephen Tyler, holder of an endowed chair of anthropology at Rice University. Professor Tyler previously had originated the scientifically oriented field of cognitive anthropology when he was at the University of California, Berkeley, but later abandoned it to be at the forefront of post-modern criticism. In his foreword, he wrote:

> Improvise *(in-pro-videre),* the un-for-seen and unprovided-for is the negation of foresight, of planned-for, of doing provided for by knowing, and of the control of the past over the present and future. Doing, unguided by "how-to," and uninformed by "knowing"— those other names for the past, the already seen—makes the open-ing for an art that is neither a craft nor a technology capable of being mastered. No mystagoguery of mastery encumbers the improvident being-now, and no history in-forms it. (Keeney, 1990, p. x)

Creative therapists need to invent individual styles that grow out of their own ways of being and avoid "the easy tricks, the unthinking rule-following, the imitation-of-the-master, the ready-made recipes of cults and charismatics, the lust for power" (Keeney, 1990, p. xi). Therapy as a per-forming art requires our avoidance of purposeful repetition and instead a willingness to play with the tune, tempo, and themes so as to per-form rather than in-form. Tyler says of this change in the contextualization of therapy: "A therapeutic encounter becomes a conversation in which thera-pist and client respond to one another without benefit of a script or even a narrative." We must suspend our habits of trying to over-control the way things should go and allow space for a surprising new direction to appear. As Tyler argued, this form of performance art is nothing less than the des-perately needed ongoing "therapy of psychotherapy."

After moving the context of therapy into the performing arts, I next took on flipping therapy out of handling problems and solutions to working and playing with the natural resources that emerge in a conversation. With my colleague Professor Wendel Ray of the University of Louisiana at Monroe, I developed a distinct orientation called *Resource Focused Therapy* (Ray & Keeney, 1993) that was neither problem nor solution focused. Because problems imply solutions and vice versa, a shift from one side of that whole interaction was nothing more than a first-order change, that is, more of the same kind of therapy. Leaving both problems and solutions represented a higher order change. To some people's surprise, this change in outlook appeared to entail a walk straight out of therapy, with an absence of therapy clichés and patter. In speaking less like a therapist, one becomes more therapeutic.

Soon after these context-shifting developments for therapy, I took a sabbatical to teach in southern Africa. While there, I began fieldwork with indigenous cultures, observing first-hand how their ways of addressing the issues of everyday living could teach us more about therapy as a performing art. My originally planned year-long sabbatical turned into over a decade of fieldwork. With foundation funding, I was able to study cultural healing traditions throughout the world, from Africa to South America, Central America, Australia, Japan, Bali, and the Caribbean, among other places.

Though I disappeared from the Euro-American clinical teaching scene, I still taught therapy in other foreign universities and institutes. I authored a book series, an encyclopedia on some of the world's healing traditions, entitled *Profiles of Healing*. One of the books was a biographical account of Milton H. Erickson, cowritten with his daughter, Betty Alice Erickson (Erickson & Keeney, 2006). Gaining a global perspective on the healing arts enabled me to appreciate that our therapy profession, as sophisticated as we may think we are, is one of the world's youngest folk healing traditions. Compare its infancy to the healing practice of the Kalahari Bushmen, which has been depicted on ancient rock art images and is still practiced today. Their healing orientation has been estimated to be as old as 60,000 years. As I explained to my biographers, Jeffrey Kottler and Jon Carlson (2004), we have much to learn from our ancestral cultures when it comes to the accrued wisdom associated with the performance of healing interaction.

The elder traditions remind us that it is possible to become a wise and effective therapist (or healer) without knowing how to read or write. Creative transformation takes place in cultures that have no professional licensing exams, graduate degrees, or professional libraries. Our oldest ancestral cultures do not need lofty technical talk to convince one another that they are really doing something. Their work is typically open and available to being experienced by the whole community. Anyone can see whether it is creatively alive. Contrast this to some renowned contemporary therapists whose performances are kept in the dark—never witnessed, only summarized, narrated, or theoretically discussed.

Most importantly, the oldest cultures regard the *whole community* as both the provider and recipient of healing. When someone is "sick," the whole community gathers and everyone is doctored and everyone contributes to doctoring. The oldest therapeutic traditions are contextual, systemic, and holistic. They use everything that facilitates creative expression, from music to dance to theatrical staging. They do not get as easily lost in abstraction, are less attached to talking about therapy, and see healing as an opportunity for becoming more fully alive. They are our original creative therapists.

I left fieldwork to return to teaching and practicing therapy. I teach creative thinking and inquiry at the doctoral program in transformative studies at the California Institute of Integral Studies in San Francisco. I conduct clinical work with Delta families living in shanties near the Mississippi River in northeastern Louisiana, described by *Time* magazine as the poorest place in America (White, 1997). In addition, I see adolescents incarcerated for committing violent crimes as well as so-called "difficult cases" that are presented in university and teaching clinics from Louisiana and Arkansas to Brazil, Africa, and Japan, among other places. I have intensively studied what is creatively possible in the performing art of therapy. Here, I will present some exemplary cases through written narratives, transcriptions, and the attached DVD.

This work serves to liberate therapy to be more creative and transformative. My aim is nothing less than the recontextualization of therapy as a creative improvisational art that emphasizes people's resources rather than adding more rhetoric that highlights, underscores, gives more detail, or invents explanations and understandings of impoverishing experiences. We do the latter when we use psychiatric diagnoses, explore history to substantiate why people are ill, articulate narratives that are viewed as needing social editing, look for instances of solutions that paradoxically maintain the presence of the problems they are interlinked with, or bicker over which interpretation is the top dog(ma).

As I will show, therapy must look, sound, and feel less like therapy to be more therapeutic and healing. Clinical work deserves to be recognized as a legitimate performing art that is as requiring of talent and skill as is the production of a theatrical play. Whether therapy markets itself as analytic, postmodern, behavioral, interactional, systemic, narrative, feminist, positive, constructionist, constructivist, experiential, scientific, outcome-based, or neurolinguistic is less important than whether it ascends into the realm of being a dynamic transformative art. The latter includes any and all of the contributions of the former domains of knowledge but expresses them only when appropriate, that is, called forth by the situation at hand.

When it takes an expert to explain that a particular clinical conversation constitutes therapy, it arguably is not a situation involving transformation. When therapy is alive, anyone can sense it. An observer off the street can see, hear, and feel the presence of a transformative encounter. It does not take theoretical discourse to argue that something happened. Transformative creative therapy, like any other transformative art, is obvious to everyone, even to uneducated and uninitiated outsiders. (However, meaningful scholarly analysis may later enable us to discern the "fingerprints of transformation," as Don Arispe [personal communication, November 6, 2008] suggests.)

As a social experiment and teaching exercise, I once showed a videotape of a famous psychotherapist to a group of undergraduate college students. I told them nothing about the therapist but asked them to determine whether the observed clinician was skilled. I invited them to articulate whatever they observed. The students readily noticed how the encounter was neither alive, nor therapeutic, nor transformative. At the next class, I repeated the experiment again, this time showing two other world-renowned therapists. Again, all the students could see in each case was the impotent presence of someone who supposedly did not know how to help awaken or move a stuck situation.

I submit this finding as an assessment of the entire therapy profession. Too often, its leaders and teachers are creatively impotent, boring, lost, or existentially dead in their sessions. With Orwellian logic, some even claim that the essence of good therapy is being boring! Even worse, some clinical practitioners can be demeaning, pathological, disrespectful, inappropriately controlling, and producers of outright miserable theatre. In spite of these professional shortcomings, there are therapists hidden in small town offices and big city clinics who sometimes have the courage to ignore the so-called masters, workshop gurus, psychotherapy bestsellers, academic textbooks, and formal education and decide to wing it unassisted by professional knowledge. With curiosity and ingenuity, they sometimes find their way into the performing art of transformative, creative therapeutic encounter. It is to these outcast performers of creative therapy that this book pays tribute. They are the best hope for the future vitality of our young healing tradition. Perhaps it is time for the members of an ailing profession to pay attention to and learn from the mavericks and outsiders, the unknown "outcast performers" who, in their exile, have taken a stand against the nonsense.

There is another part of my social experiment and teaching exercise that was conducted in a small liberal arts college in Santa Fe, New Mexico. I invited the students to invent ways of performing healing transformation. They created dramatic rituals for one another, including a grand procession of a young woman's never-before-seen painting (because she was too shy to reveal her aesthetic work) through the campus that was followed by an all-night drumming and dance party.

During that same semester, Katrina smashed into New Orleans and all but destroyed the spirit of a great American city. The students decided to create a mini-French Quarter in an outside yard located between some old dilapidated buildings. Walls were painted, musical instruments and Mardi Gras beads were hung from trees, a stage was built, and finally, a phone call was made by the class to some musicians who, before the disaster, had lived in the Lower Ninth Ward of New Orleans, the area most damaged by the flooding. The musicians were invited to come to Santa Fe, where they were told that a home for the "spirit of New Orleans" had been built. The performers from New Orleans came and played a heartfelt concert where students and musicians danced, feasted, laughed, and wept, while planting new seeds of hope for everyone's future.

I extend the same invitation to you. Take off your comfortable therapy glasses and see the world of therapy as a performing art. Build your own stage and bring your own improvisational style. Have the courage to be part of the newborn hope that is possible when we pack up our therapeutic belongings and head for the performing arts. We need a revitalized theatre for the performance of this transformative art.

Gather your instruments and talents, whatever they may be, and get ready for an ongoing therapy of therapy. Even if you do not dance, sing, or sculpt, you have something to bring to the table. Fishing or cooking will work as well. Yes, I have been with a family that cooked fried chicken in their kitchen, but we creatively did so with transformative good taste. Furthermore, I have fished with my clients—where the fishing pole was in my office! Expect therapy to become the most thrilling and exhilarating (and often unexpected) performing art you have ever experienced. Lights on!

PART I

Bringing It Forth

CHAPTER 1

Theatre of Creative Transformation

Setting Up the Three-Act Play

All schools of therapy are vulnerable to getting lost in a session (which is not always a bad thing!) because they lack a compass or a map—they have no practical way of keeping track of their location and movement. Clinicians sometimes don't know whether a session has even moved one step forward. Similarly, they may be unaware when it has backed up and become more entrenched in a stuck place or gone down a side road where everyone is completely lost in a different form of entangled discourse. How can we keep track of where we are in the flow of a session?

Most of the arts have a practical means of scoring the movement of performance. For example, a musical score tracks the melodic line (along with harmony and rhythm), whereas dance notation symbolically lays out the dance moves. With a musical score, a musician is aware of where he or she is in the song—where the notes of the melody are in relation to the beginning and end, whether it is the verse or chorus, and so forth. A clinician without a means of scoring his or her journey is like a musician without a song (whether internalized or written), or a dancer without choreography, or a filmmaker without a screenplay. It's probably even worse—it's more like an explorer without a compass and map.

Of course there are times when performers spontaneously make up a song, dance, or play without following a score. But at any time during the flow of the creation, or afterward, they know how to map the movement—the progression of notes, steps, and themes they went through. Therapists generally don't know how to do this. They typically haven't even asked the question. They just make theoretical generalizations about what they think happened on an abstract level. This is probably because the profession has lost awareness of its being primarily a *performing art*. It therefore has grown accustomed to giving less importance to what is actually taking place in a session—its *live performance*—than to indulging in the hermeneutics of interpreting it.

The elementary three-act structure of a theatrical play or movie screenplay provides a starting point for helping us know where we are in a clinical session. All performance, including the clinical theatre of therapy, can be scored so as to notate the progression through a beginning, middle, and end. Keep in mind that the end may be another beginning or the middle (or start) may be experienced as some kind of end. The movement does not have to be linear. A session might turn out to be a big U-turn, or a running in circles, or even a progressive spiral, to mention a few possible patterns.

Syd Field (1979), the author of *Screenplay: The Foundations of Screenwriting,* outlines the pattern that most screenplays follow. They usually work with a three-act outline that marks the temporal flow of a performance in terms of beginning, middle, and end acts. The transitions between acts are called *plot points,* those moments or events that shift the plot in a new direction. As an elementary form for the staging of theatre and film, the basic structure for the performance of creative therapy also can be based on the outline of the three-part play. It gives us a practical way to keep track of where we are in a session and whether we are stuck or have moved in any direction.

Again, all performed plays benefit from having a means of knowing whether they are in the beginning, middle, or end. This also applies to the producer of a therapeutic play. If you end a session and haven't moved anywhere, you need to be aware that you got stuck in the opening act. If you find yourself in the middle, it's good to know that you moved forward but that there was no achievement of resolution or closure. Therapists of all schools and orientations can learn from the lessons of screenwriting how to know where they are in the unfolding of a therapeutic plot, storyline, or dramatic enactment.

Creative therapy aims to bring forth a session that is well formed and whole. That is, every time you meet with clients, you aim to have a beginning, middle, and end. Each session should hold the progression of all three acts, that is, be a whole session. This approach used to be called *single session therapy*. Let's now call it *whole therapy*. To illustrate:

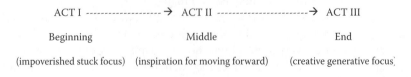

Figure 1.1 Three-act structure of a therapeutic performance.

In the beginning, people usually (but not always) complain. They are stuck in an onslaught of complaints. The more they and others think, talk, or do anything about this impoverished focus, the more out of focus everything becomes. Act I is like quicksand. Most therapists contribute to helping clients sink deeper into this stuck place. They bring forth more information, provide alternative explanations, pose new questions, edit impoverished stories, ponder miracles, or attempt all kinds and orders of solutions. This approach helps keep everyone in Act I, potentially for the rest of the client's life. Act I is a potential "burnout zone," where therapists are most likely to think they have the most frustrating or boring job in the world.

The creative therapist waits for any inspiration, sign, imaginative leap, clue, or opening that points the way out of Act I. It doesn't matter if the possible direction for movement is rational or irrational, enlightening or absurd, relevant or irrelevant. Any creative inspiration that might carry the therapist and client to a middle act is utilized. The therapist must go past his or her conditioning and have the courage to walk on an unknown road. This is when movement is felt for the first time. At this point, the therapist does everything possible to keep the session hanging in the interface between beginning and end, despair and hope, old and new (as opposed to falling back into the beginning). Here, chaos and craziness may make a house call to help nourish the process of transformation. This is where therapists think they have the most irrational job in the world.

Act II is the fulcrum—things can fall backward or forward. The more it teeters, the more likely the previously stuck frame will be loosened. As Archimedes once said: "Give me a lever long enough and a fulcrum on which to place it, and I shall move a world." The creative therapist must move far enough away from Act I and stand on a fulcrum that enables the client's paralyzed world to move.

Finally, the transformative leverage brought forth by creative interaction moves both client and therapist to Act III. Here, they find themselves in a new experiential reality where silver creative linings, rather than oppressive therapy clouds that call them to hide in a psychological shelter, are witnessed. Without effort, the sun shines in this act, even when it is dark. Creativity is in charge. Clients and therapists are surprised, often with a sense of awe, at how alive and generative their sessions have become. The process no longer feels like therapy; it has become a sizzling *theatre of creative transformation*. New possibilities, considerations, and scenarios for daily action are brought to light and initiated. This is when therapists think that they have the best job in the world.

That's the whole of it. Everything else is mere detail.

The Creative Turn

Let me present this orientation in a slightly different way: Act I presents what the client brings to a session, "the presenting communications." This act is the set-up, the beginning of a session, typically voiced as descriptions of problems, symptoms, and suffering. At this point, therapists need to avoid wallowing in the obvious and instead follow the advice of naturalist author Cathy Johnson (in Barron, Montuori, & Barron, 1997), who extols the importance of wandering and getting lost in a search for unknown experience. "Wandering is the itinerary" and it requires "a willingness to go beyond my safe, homey environment, my comfortable and comforting preconceptions" (p. 60). Wandering involves our being "unprogrammed" and open to finding an unexpected treasure lying just beyond our peripheral vision.

Act II moves the session to address a theme or topic that was completely unexpected and that, on the surface, seems to have little or nothing to do with the presenting communication. Inspired by an accidental encounter with the unknown, this is where the possibility for a *creative turn* takes place. As biologist Karry Mullis (in Barron et al., 1997) puts it, find what "you don't know anything about and look at it for a long time, and you might learn something totally different" (p. 73). Attend to what the clients and you would otherwise bypass because it lies outside of problem or solution talk or outside of habituated meanings. Consider it an exit from therapy and a pointer toward possible transformation.

In Act II, the therapist holds onto the presence of something new and unexpected and works with it—both keeping it as a main theme and trying to amplify the intensity and validity of its presence. Act III takes place after therapy has successfully managed to creatively turn the theme and focus of the session into a "creative zone" that calls forth even more spontaneous creative moments. In this Part, there is no effort; everyone, clients and therapists alike, are in "the zone." Anne Dillard (in Barron et al., 1997) captures this situation well: "Beauty and grace are performed whether or not we will or sense them. The least we can do is try to be there" (p. 84).

Of course, the simple three-act structure may give way to more complexity. More acts are possible and they may involve unanticipated forms of movement. Mapping or scoring the acts of a session, no matter what form they traverse, helps us grasp the overall flow of a case as it is happening in real time or post hoc analysis. Like a musical score, scored therapy sessions show whether a conversation is moving (or not moving) from one act to another. Therapy that has been creatively awakened will move somewhere, as if having a life of its own, in the same way that an effective melody, story, screenplay, or theatrical play enacts movement.

Getting the Soil Ready with Good Timing

Peter Ilyich Tchaikovsky (in Barron et al., 1997) had this to say about creativity:

> If the soil is ready—that is to say, if the disposition for work is there—[the seed] takes root with extraordinary force and rapidity, shoots up through the earth, puts forth branches, leaves and, finally, blossoms. I cannot define the creative process in any other way than by this simile. The great difficulty is that the germ must appear at a favourable moment, the rest goes of itself. It would be vain of me to try to put into words that immeasurable sense of bliss which comes over me directly [when] a new idea awakens in me and begins to assume a definite form. I forget everything and behave like a madman. Everything within me starts pulsing and quivering; hardly have I begun the sketch ere one thought follows another. (pp. 180–181)

Many years later, Frank Zappa (in Barron et al., 1997, p. 197) added his advice for how to compose. I have transposed it for therapy:

1. Declare your intent to create a session.
2. Start the session at some time.
3. Help something happen over a period of time.
4. End the session at some time.
5. Get a part-time job so you can always continue doing sessions like this.

Tchaikovsky and Zappa provide another perspective on what we need to know in order to create transformative artistic work. The challenge is in finding out how to ready the soil, recognize the favorable moment, and know when to end. In other words, we need good timing and rhythm. Duke Ellington adds one more thing: "It don't mean a thing if it ain't got that swing." Syncopation must be added to make the beat swing and that, in turn, gives life to music. The art of therapy similarly requires uplifting a session with rhythms that are capable of bringing forth creative movement.

To recapitulate, time in therapy most generally refers to moving through the beginning, middle, and end of each composition. Act I is the soil that must be prepared so that the ecology is ready to support new growth. In this stage, the previous dark nights of a client's soul are utilized as tillage and compost, providing fertile soil for entry into another chapter of life. Act II introduces and plants a seed into the previously prepared therapeutic soil. This seed must receive appropriate attention. It requires watering and nutrition so that it may germinate and plant its roots deep into the dark soil. It initiates growth. Act III is harvest time. The seed has broken through the ground and is now fully in the light. It reaches upward and finds an awakening of fully embodied presence in the world. This stage helps bring forth the ripe fruit of therapeutic transformation.

Time in therapy also refers to the tempo and beat to which its movement is paced. Suffice it to say that the art of therapy has as much to do with timing, tempo, and vibrant rhythm as anything else. The therapist must move along with the natural unfolding of a session and be ready to accompany, support, and encourage its progression with inspiring rhythm, like a percussionist in a jazz trio. The "beat" applies not only to music but also to therapy. Without good rhythm, therapy can easily get off track, become disjointed, and lose its momentum. When the timing is alive, we can say that a session breathes and moves with its own spirited pulse.

Timing gives us another way of understanding change and transformation. It asks us to be entrained with our client, that is, rhythmically coordinated. This is another way of saying that we need to be in step with one another and swept away by the dance. It is the rhythm that grabs us, dances us, and moves us somewhere. This is the felt (as opposed to abstracted) "soul" of therapy.

I once met an old Lakota medicine man famous for presumably knowing how to make it rain. When I asked him how this was possible, he answered, "You have to be at the right place at the right time." Being a rainmaker in therapy is the same: make sure you have good timing and are in step with the natural processes of change already surrounding you.

The execution of therapeutic creativity requires knowing when to introduce something, when to move on to the next thing, when to move from one act to another and, most important, when to end a session and send the clients back to their daily life. As a clinical supervisor, I have seen many sessions undo themselves by not ending when they should have. When a session is over, stop. If a session is well formed (it has gone through all three acts) after only 5 minutes, then get up and end it. Many of my sessions last between 20 and 30 minutes. They may be less or they may be more, depending on the song played, the dance danced, or the improv that has unfolded. It's all in the rhythm.

Surprise Ending: Sometimes Paradigms Ain't Worth a Dime

We should not forget that the three-act structure of a simple play was simply invented. Though called a "paradigm" for the construction of screenplays, once prescribed, it becomes what Geuens (2000, p. 107) calls a "super genre" that too easily dictates the only acceptable form for unfolding a play, film, or story. When asked about the traditional three-act paradigm, award-winning producer and screenwriter Diana Osberg (2006) responded, "Beginning screenwriters should learn the three-act structure thoroughly before they start breaking the rules" (p. 1). However, she qualifies this:

"It appears that we human beings respond to the three-act structure in a primitive way. This 'arrangement' connects with us at a deep, subconscious level that gets us into the theatre again and again."

By convention, most plays, movies, and stories first set up the circumstances, main characters, time and place, and conditions as a starting point. The audience knows that a plot has begun. Then a plot point appears—something catches our interest, like a specific unexpected event that the main character must address. We are drawn into the movement of the plot line. Entering the second act, our protagonist often faces all kinds of challenges and obstacles before another surprising plot point opens the next door to the final act. There, matters are spun around, enabling our character to achieve (or resolve in some way) the goal of reaching an ending.

From fairy tales to hero and heroine adventures and lost and found journeys, this structure has traversed the well-worn road from "once upon a time" to "they lived happily ever after." As has been repeatedly suggested, life may not follow such a progression of staged acts and chapters. Obviously, the clients who visit us are typically stuck in a scene they seem unable to escape. From a conventional playwright's perspective, they need a plot point, a creative turn that can deliver them somewhere else.

I first introduced the idea of the sequenced progression of frames, or what appeared as narrative structure, to the field of psychotherapy in 1983, with the outlined progression of themes (frames) in a case study of Olga Silverstein (Keeney, Ross, & Silverstein, 1983). In the mid-80s, I began using the three-act structure of plays as a practical means of scoring therapeutic conversation. During that time, as professor and program director of the Texas Tech University family therapy doctoral program, my colleagues and I organized, discussed, and analyzed cases through an elaborated scoring method I developed called "recursive frame analysis." The earliest version of this form of analysis was presented in 1985, as the core of a book written with my colleague, Jeffrey Ross, *Mind in Therapy: Constructing Systemic Family Therapies*. The fully elaborated method of scoring cases as conversational movement through scenes in a session was published in 1990 in the journal *Terapia Familiare*.

Prior to 1988, I presented this fully ripened and practiced work to various family therapy institutes, including several venues in Adelaide, Australia, where Michael White attended my workshops. There, I made the argument that therapy should be contextualized as a performing art akin to theatre. Furthermore, I proposed that it would be tempting, but erroneous, to frame my work as a "narrative therapy" (this was before that name had been used to describe any clinical work). I argued that an emphasis upon "narrative" would miss capturing the richness of the live performance of therapy as theatre. Finally, I argued that drama and theatre were more suitable metaphors than narrative and the telling of story.

Plays are performed live by the main actors, rather than remembered stories told by a narrator. Action is the main ingredient of a play and, inside the theatre, dialogue is considered as action. In the telling of a story, there are simply words. Though words can be fantasized as externalized and suspended in time for post hoc editing, thickening, thinning, resorting, reprocessing, restructuring, or retelling, they remove us from present state interaction. In contrast, being inside the ongoing live performance of a drama, rather than telling a story, enables us to be present in the actual here and now. Stories distract us from being fully present in favor of examining what occurred in the past, though they may project a future (derived from the storied past). They are always an historical fiction, whereas a live performance is the actuality itself. The bringing forth of action—rather than having action recalled and described—is embodied by the live performance of therapy.

Looking back, I am grateful that I was never tempted to contextualize therapy as narrative work. The field of psychotherapy long before had shifted to highlighting performance with the unique contributions of so many pioneer performance-oriented therapists with their here-and-now interactions, both in psychotherapy and family therapy. I am grateful that some of them were my first teachers and mentors. Engaged, improvised (moved by the moment), creative performance was the transformative difference, rather than any shift in theoretical understanding, interpretation, or story. More accurately, an emphasis on live performance ensures that all interpretations will never stop shifting and moving, unable to maintain a singular stronghold on a therapist or the field. The dramas of therapy are not limited to being *psycho*dramas or family dramas, or any framing of drama. The theatre of creative transformation welcomes whatever is brought forth and improvised.

Surprise Middle Act: The Primacy of Experience

What I am saying is that the major breakthrough in the therapeutic arts took place when therapists embodied and highlighted *performance* rather than narration and interpretation. This breakthrough is lost when clinicians remove themselves from the unknown, unpredictable, unscripted, and necessarily improvised nature of human encounters. When Carl Whitaker made light of theory and called it a "hindrance" to clinical work, he was inviting us to perform rather than narrate and interpret. His call was for the primacy of *experience* in therapy.

Unfortunately, an emphasis upon interpretation, narration, and commentary quiets, darkens, and deadens the stages of live performance. With this approach, therapists and clients are tempted to limit themselves to telling stories or offering reflections without being interactionally mindful that they are live on stage as they are doing it (and getting deader by the moment). Therapy then regresses back to where it had been in its beginning days, when it asked clients to get off their feet and lie down on a couch. In the horizontal plane, the linear lines of storytelling more easily take precedence over danced interaction. The therapist retreats into an "outside external position" of observing/telling/witnessing *about* lives, rather than being on a live stage that calls forth change in the moment in which it is performed.

The position of narrator, whether assumed by client or therapist, removes one from the live stage and exalts the role of the audience over the actors. Therapists and clients externalize one another and become disembodied at the cost of losing their performed therapeutic interaction. Compare this to the electrical-like tingling experience of spontaneous, noninterpreted, embodied transformation of live performance. It is something you never get over. Martha Graham describes it this way: "There is a vitality, a life force, an energy, a quickening, that is translated through you into action, and because there is only one of you in all time, this expression is unique" (de Mille, 1991, p. 264). Once you have felt this aliveness, you can never be content to sit on the sidelines offering interpretations or reflections.

However, as witnessed from an outside observing position, it must look next to impossible to *know* how to jump into the creative current. The "be spontaneous" paradox stymies the outsider—usually voiced as, "What do I need to know, understand, interpret, narrate, believe in order to become a part of the transformation?" This question will cognitively strap you down and disable entry into performed action. There is no knowing that directs how to be creatively alive. You must jump into the flux and have the greater unknowable whole, what we sometimes call "complexity," sort you out and carry you along.

Because most *writers* (and *speakers*) are more comfortable and accustomed to being outside observers and commentators, they presumably never emphasized or even noticed that the *dynamic performances* of Milton Erickson, Virginia Satir, Don Jackson, Salvador Minuchin, Olga Silverstein, Peggy Papp, Frieda Fromm-Reichmann, Cloe Madanes, and Carl Whitaker, among others, had less to do with the theories they were articulating after a session than with the way they were actively engaged with clients in a session. Writing, the post-hoc produced commentary of an outside observer, arguably does not facilitate making entry into performance. It perpetuates the observing, listening, reflecting, commenting, narrating posture.

One of the most respected narrators and commentators of the family therapy field, my early colleague Lynn Hoffman, established her early reputation as a skillful reporter about the founding pioneers of family therapy. As I look back at this history, she may have tragically missed the significance of *active interactive engagement* and instead, later in her career, overindulged in contemplation about which *interpretations* were most appealing or "correct." When this happens, the history of a field is depicted as an "evolution" from one interpretation to another. In the beginning was psychodynamic interpretation, followed by behavioral interpretation, phenomenological interpretation, systemic interpretation, feminist interpretation, narrative interpretation, postmodern interpretation, postcolonial interpretation, and on and on. This kind of historiography, from the perspective of performance, obliterates live action and renders the play unimportant. The theatres go dark and what remains are narrators and critics sitting still in the audience making interpretations. The reflecting critic, who has no need to develop any skills to perform, chooses to not act and not perform, preferring to write about doing nothing, thereby making inaction the desired action.

In the blink of an observing eye, therapy lost its creative staging of performance. In the case of family therapy, its interpreters dragged it back to a state that was more like those of the historical psychotherapies from which it had previously differentiated itself. The field became competitively organized around whose interpretation was more correct. It too often displayed a noncollaborative meeting of egotists—an "I for an I" rather than an invitation to act and interact in relationship. The profession became blind (and blind to its blindness) to seeing interactive therapeutic performance.

I recognize that the sociology of a profession is organized by politics—personal, institutional, and cultural. Yes, along with everything else, there were unheard voices and missing presences. Without doubt, white male doctors exercised too much decision making. The whole therapy culture arrogantly advanced games of one-ups*man*ship over relational caring and ecological nurturance. God help us if the behind-the-scenes stories of therapy's politics were ever disclosed.

Not surprisingly, the articulation of problematic concerns—from ethics to aesthetics, gender to culture, and private practice to professional guild—took place through the same mediums that had perpetuated the inequalities, discriminations, and pathologies the field previously rallied against. Rather than offering alternative contexts, mediums, forms of presence, and performance, we ended up with more noninteractive monologues, musical chair–like reorderings of the same hierarchies, and more (rather than less) interpretation and narration with new voices of privilege. What we desperately needed, then and now, is the performance, enactment, and embodiment of a more diverse, polyphonic, relational way of *being in* (rather than separate from) the whole of therapeutic performance.

In the area of ethnicity and culture, therapists typically did not bother to experience how other cultures *perform* their ways of healing and transformative engagement. Instead, *academic interpretations* of diversity were offered, most of which were derived from the privileged discourse of sociological stereotyping and categorization. An important paper entitled "Family Therapy and Anthropology," written by Tullio Maranhão (1984, p. 267), denounced a seminal book in the field, *Ethnicity and Family Therapy* (McGoldrick, Pearce, & Giordano, 1982), saying it, "reads as a sort of Guinness Book of Cultural Stereotypes." Maranhão, a Brazilian anthropologist with a doctorate from Harvard, a faculty member at Rice University at the time of the article, and an internationally leading scholar on dialogue and postmodern thinking, argued that "the list of cultural traits [provided by the book] indicates exactly those kinds of shallow prejudices about groups of nationals, prejudices that should be avoided especially by family therapists" (p. 268).

Though well intended, family therapy's early efforts to identify generalized cultural descriptions and associate them with therapeutic prescriptions too easily sidestepped the more sensitive interaction of being with each family as if it constituted its own unique culture that required a therapeutic presence called forth by the relational interactive dance of the moment. When performance is minimized and interpretation highlighted, we inevitably end up with wide-sweeping generalities that sometimes stray off mark and become the very thing we are trying to remedy.

We need more presence of "diverse cultural ways" that are voiced through their own modes of expression. We will know this when we attend symposiums that are more than shout fests but rather invite a wide diversity of cultural forms of expression that may include dancing and singing For instance, the world's oldest healing tradition, held by the Kalahari Ju/'hoan Bushmen, involves "not talking" about problems. The Kalahari Ju/'hoan Bushmen dance in order to shake themselves free from the way words and narratives get human beings stuck (Keeney, 2005). Perhaps psychotherapy could benefit from a good old-fashioned Kalahari all-night healing dance. I propose that our conferences, academic programs, journals, and everyday clinical sessions start walking the talk: let's welcome diverse cultural forms of *expressive performance* rather than solely legitimizing the Euro-American mode of category-making, interpretation, narration, and explanatory text.

Issues about diversity require greater insight into the ways performance, including the performance of interpretation and criticism, may enact, confound, or negate an intended message. It is time to explore different ways of staging how differences can be communicated in a medium that does not imply privilege for any particular cultural mode. We need less interpretative criticism voiced by privileged forms, and more creative exploration of how different cultural traditions and individuals can have their own stage for their unique ways of performance. We need more than interpretative criticism voiced on a colonized stage; we need performed differences on diverse stages.

Like old-fashioned family therapists and indigenous healers, we must all learn how to be more relational in holding the interactive dance of separateness-in-relationship-with-togetherness. In the yin-yang of distinction and relation, we find common and uncommon grounds. Our cultural and gender differences, as well as other forms of difference, must be clearly distinguished in order to interact in relationship, and our unity in relationship must follow from the equal participation of each side of a distinction. We need to mother and father a way of holding one another's differences.

Feminism has asked for this relational holding. Though early voices of feminism had less to do with relational presence than with distinguishing previously unvoiced differentiation, this was arguably necessary. As G. Spencer-Brown's (1973) "laws of form" formally noted, the distinction must first be drawn before relation can follow.

Now is the time for us to evolve a "feminism of feminism," in which our profession would enact relational presence on the stage of therapy. Like all discourse that carries news of "difference," we must be careful not to overemphasize interpretations and hierarchical repositioning that may negate the very relational thinking and presence we desperately need to experience. Feminists like Rosemary Reuther (1992, n.d.) and Charlene Spretnak (1993) are writing about "ecofeminism" and turning to systemic and ecological ideas and metaphors which, after all, were always about relational thinking and interaction, though often lost in "systemic interpretation."

Getting closer to my own earlier work, advocates of systemic therapy were no less guilty of abandoning the stage in favor of observer-able interpretation. The Milan therapy quartet, originally directed by the psychoanalyst Mara Selvini-Palazzoli (1979), offered an operatic style of systemic *interpretations*. Think of the whole family lying on an oversized analytic couch, with the therapist sitting in a neutral position while asking questions, until the final grand interpretation, and you will get the picture of this style of "family systems analytical therapy." I should note, however, that their interpretations aimed to paradoxically bring forth a transformation in family interaction outside the clinic, but their sessions were not yet emancipated from a narrating interpretative perspective.

In my own contributions to family therapy in the early 1980s, I advocated an aesthetic framework for change in therapy, a relational epistemology tied to cybernetic, systems, and ecological ideas. Although my book *Aesthetics of Change* (1983) was cited in the *Encyclopedia of Artificial Intelligence* by cybernetics founder Heinz von Foerster (1987) as one of the six general references for cybernetics, it and the published papers that preceded its publication stirred up intense controversy in family therapy. In an issue of *Family Process*, I responded by saying that therapy was neither about an exclusive aesthetic or pragmatic frame (Keeney, 1982). What was it about? I alluded to the complementary dance possible between distinctions that could be transformed from dualisms to yin-yang relations. But in my heart, I felt something was wrong with everyone's discourse, including my own. Though I struggled to articulate what was off about family therapy discourse, I knew what was right about the history of family therapy and psychotherapy in general. It was found in its performances, not its interpretations.

During the height of this controversy, I left one of the citadels of family therapy in New York City and headed to the western plains of Texas, where I immersed myself in the improvisational nature of therapeutic performance. In this frontier setting, I was contextually free to fully emphasize the performance of therapy. I brought the postmodern thinker and anthropologist Stephen Tyler as an adjunct professor to the doctoral program I directed. When my close colleague Harry Goolishian, who also was an adjunct, first (mis)heard Tyler's postmodern discourse, he jumped off his previous action stage and began proselytizing postmodern interpretation. His therapy immediately became boring, as he himself observed, and he began teaching that "good therapy is boring therapy." Though I teased him about this nonsense and we maintained a mutual respect for one another (at that time, he asked me to become director of his institute), I lamented the end of the field of family therapy as a stage for *performing* creative therapy. It became lost in interpretation.

In the subsequent renaissance of interpretative muddle, the performance of creative transformation on the therapy stage was all but forgotten. Even worse, the new grand interpreters created their own hegemony, privileged hierarchy, and disqualifications of others (the very things they were against) and led a crusade against the extraordinary contributions of the previous great actors of therapeutic transformation, from Satir to Erickson to Whitaker. As I reflected on this many years later with Salvador Minuchin, it became clear to me that the field of family therapy cannibalized itself. It ate its own remarkable ancestors and spat out words, interpretations, and grand theoretical discourse, all accompanied by the absence of thriving creative action. The same has happened over and over again to other traditions of psychotherapy: creative performance is silenced for interpretative commentary.

Fortunately, interpretative therapists—whether narrative, postmodern, systemic, or analytic—find it practically impossible to remain totally and forever detached from the flow of action. In their best moments, they are honest enough to get fed up with being a bored outside interlocutor and, out of the sheer need to wake themselves up from falling asleep in a session, spontaneously find the courage to step on stage and act as the drama calls them forth.

What interpretative therapies too easily miss and frequently fail to utilize is interactive participation in the live dramas clients enact. Transformation takes place in relationship with therapists who have been recruited to dance and act with their clients rather than cautiously narrate or interpret them. Let us pledge to never forget that the heart of our art is *live performance*.

Virginia Woolf (1926) knew that there is "some secret language which we feel and see, but never speak" (p. 314) and that we should sometimes not offer to tell it because that would kill what begs to be experienced. In this regard, artistic filmmaking has historically displayed hostility toward narrative, not wanting it to cover up the empty spaces and interstices of human drama. As Nietzsche put it, "plot exists simply to occupy the front of our minds, whilst the music works on our souls" (cited by Turner, 2005, p. 1).

Sarah Turner (2005) argues that stories are "vampirical," at most providing hooks to hang images on (in the case of filmmaking) or as semantic hooks on which to hang time-frozen clients and therapists, prioritizing the content of what is told over the live telling and enactment. In live drama, the possibilities for transformation are breathed, metaphors are animated, and emotions mingle together and affect one another in an ever-shifting flux that refuses to be framed by any narrator's or interpreter's rendering.

Back in the 1980s, I was cautious about the "story line," believing that it was "recursive shape-shifting circles" rather than "hegemonic static lines" that were needed. At that time, I purposively embedded each scene of the so-called acts of a play inside one another to show them as recursively rather than lineally structured. From the perspective of nonlinear time, or what H. Maturana (personal communication, October 7, 2008) has called "zero-time cybernetics," there is no lineal progression. Yet, I also knew that as important as the perspective of the "simultaneity of interactions" may be, it is still necessary to recognize that the constructions of time and line can serve as pragmatic approximations of more encompassing complexity. In this regard, it is sometimes convenient for a construction worker to use a flat-earth hypothesis when building a house or tennis court and for a playwright to use the plot line as a vehicle for structuring the unfolding of a play.

We must remember that a three-act play or story is not the whole story and it is not what is "alive" in the performance. The progression of themes and words require tone, music, rhythm, coloring, nuance, rich complexity, and recycling, among other unspoken ingredients, in order to bring it to life. The map is not the territory, the transcribed notes are not the music, the lyric is not the song, the meaning is not the lived drama, the told story is not the enacted play, the interpreted life is not life.

Being alive in a session—in the sense of experiencing the natural flow of creative expression—has to do with being inside the mind of therapy, on stage with clients, and in the mix where efforts to transform are brought forth by the sincere authentic requests of clients who have come to be changed. Fritz Perls, Virginia Perls, Peggy Papp, Cloe Madanes, and Carl Whitaker, among others, had one thing in common: they were in the action scene, part of the observed-and-heard-and-felt, and active constructors of the clinical reality. They were not outside the *experience* of transformation. They were masterful performers and evocators of change.

Having a client lie down on a couch, or on a Procrustean bed of story or theory, is most likely to deaden the action and the scene. I invite you to reconsider acting and performing so as to bring forth life in a session. Many therapists are already partway there. Although performance is somewhat about stories and scripts, it is more about the interactions held by a conversation voiced inside a whole communicational matrix. It is even more about what cannot be told. Now step on the stage, get in the action, and embody the unfolding of each unscripted, narrator-less play.

Surprise Second Beginning

Let's cautiously return to the simple structure of a three-act play, knowing that it does not hold what is essential about therapy. It is more akin to what musicians call a "cheat sheet," a short outline of the themes and their movements that enables us to keep track of where we have been, where we are, and where we might go. That's about all the three-act paradigm is, a cheat sheet. Music, theatre, dance, and therapy can move in many directions and have plots within plots that may morph in any way. A scored cheat sheet is only one prop on the stage. It may encourage or inspire the next move, but it is not the play.

With respect to the production of externalized commentary, it is important to reveal the performance that underlies the delivery of any grand-scale interpretation. Does the narrator's performance enact that which it claims to address or does it pull us away from its validation? The idea that clients are organized by interpreted stories, distinct from the performance of them, is simply an example of one therapist's story about the presence or absence of his or her performance in therapy. The notion that a postmodern orientation is "correct" therapy is another performance of a modernist polemic. When not relationally performed, feminism loses the essence of the message it was born to deliver. If not enacting the circularities of relationship, cybernetics, systems, and ecological theory lose their recursive presence. Interpreting culture without legitimizing diverse cultural expression in its performance is simply more recycled privileged discourse.

Let us free ourselves and return to the stage of unpredictable performance. There, we may be inspired to think or do anything as the situation calls it forth from moment to moment. On stage, we can be externally described as performing as a feminist at one moment, a postmodernist at another, and then a structural family therapist, a psychoanalyst, a tea leaf reader, or anything imaginable or unimaginable, often without *knowing, narrating,* or *interpreting* it.

I am not advocating an integral theory or an eclectic therapist—both of those ideas are as theoretically limited as the limited theories they purport to contain. Even worse, something is lost in mixing and integrating distinctions in the same way that colonization results in the homogenized imprint of the colonizer. I am speaking of theory-less action, performance not consciously directed, though unconsciously influenced, by knowing. I am talking about the unsaid, what cannot be said, and what should not be said. To our textual chagrin, the unspeakable unknown is the most important stage for performed therapy.

We should respect the vast ecology of perspectives. We need a plentitude of diverse knowing that includes a diversity of not knowing to enrich our therapeutic presence in the world. We do not need the hermeneutic hegemony of a theoretical colonization of any integration that accepts everything as partially true, as pieces that must be appropriately handled and fit together to make a privileged jigsaw puzzle. A complex ecology needs negations as well as affirmations, along with both inconsistencies and consistencies, logics and illogics, either/ors and yin/yangs, and the politically incorrect and politically correct, vulgar and pure, profane and sacred— as well as views that say the opposite or something slightly or radically different than this same proposal. We need not another monologue, but a live enactment, a performance and an embodiment of complex, rather than trivial, diversity. When and how our performed lines are voiced and embodied, however, need to be brought forth by the inspirations of the improvised moment. All the many ways we can frame and communicate with one another should be resources at hand, ready to spring into action, rather than monolithic totalizing ideologies that propagandize another fundamentalism of either/or zero-sum interaction.

Perhaps it would be less constrictive and more constructive to say that we are more organized by the music of our lives, or the rhythms of living, but it is even more than this concept. We live in an infinite flux of *complexus,* the multiple netlike weavings of contexts, sub-contexts, and meta-contexts. When it appears stacked, it is even more complex than Chinese boxes, where any part can suddenly become a whole at any given moment's notice. Whatever you think it is, it changes at that very instant because every particular knowing is the consequence of an interaction that, in turn, changes both knower and known. It's more complex than you thought and it's even more complex than that because it, as well as our relationship to it, just changed again.

The fluidity of experience points the way out. Rather than row to a spectator's shore and stand still talking about the flowing rivers of therapy, jump in and get carried down the flowing stream. Forget understanding; it's not so much about understanding as we have been led to understand. More importantly, respect *complexus,* complex complexity—therapy is beyond totalized understanding. Again, the alternative is to get inside the flux of therapy. The flux will move you. Feel free to listen or not listen to stories, half-stories, or antistories as you float along. But know that whatever is said holds no more importance than the silences between each telling or the rhythms that breathe back and forth between sound and silence, action and reflection, presence and absence, change and stability, beginning and end, life and death.

We need to be something other than grown men and women in therapy. We also need to be children, or at least embody the playfulness and willingness to act foolishly that characterizes the young at heart. How many therapists and therapies advocate having the courage to be a fool? How many therapists portray only extreme seriousness? Performance demands that we take risks, including the risk of making a fool of ourselves. As playwright Cynthia Heimel ("Lower Manhattan Survival Tactics," *Village Voice* Nov. 13, 1993, p.26) has suggested, "When in doubt, make a fool of yourself. There is a microscopically thin line between being brilliantly creative and acting like the most gigantic idiot on earth. So what the hell, leap."

Performed, rather than interpreted, therapy cannot be reduced to any explanatory metaphor or pragmatic schema that will help a therapist know what to express. No matter how hard you may try, you cannot jump out of the scene and avoid performing. Listening is an action and every effort not to influence, communicate, structure, or intervene becomes a paradoxical way of poorly doing the very thing you resist enacting.

Tell clients to begin anywhere, at any beginning, and then join them in a movement, any movement, toward where the flux carries you. Keep on going until you have experienced the room, the session, and the performance give birth to the life of a new home for transforming experience. Then pretend to stop. If everyone did their job, they will be captivated enough to hang out in a new way of performing their life—at least doing so until it is time for life to move things along again.

Begin the first act by acting in any way that aims to transform the experience of the clients who have come asking and paying you to help facilitate a change. When a change occurs, utilize its movements to perpetuate ongoing change of whatever is brought forth. This momentum, like a snowball getting larger and larger, will roll you into a finale where movement is rippling through the conversation, inventing new possibilities for altered action and accelerating readiness for a different world—the flux of a "changing life"—to be faced when the clinic's door is opened for everyone's exit. Consider the whole of your session as sitting in the green room, getting actors ready to go on stage to perform a new life. Pamela Travers (in Barron et al., 1997), best known as the creator of Mary Poppins, in a short story entitled "The Interviewer," has this to say: "But nothing in life—nor, perhaps in death—is ever really finished. A book for instance, is no book at all, unless, when we come to the last page, it goes on and on within us" (pp. 41–42). The same is true for whole therapy.

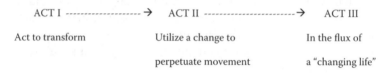

Figure 1.2 Three-act movement of creative therapeutic transformation.

Therapy is both mothering and fathering, something both men and women of all cultures are capable of fostering. We grow our clients and ourselves through growing our conversations, interactions, and spontaneous engagement. The historic experiment of Richard Held and Alan Hein (1963) demonstrated that newborn kittens that sit still grow up blind. It requires action in order to see. Similarly, we must act on the stage of creative transformation in order to bring forth awakened and illumined sessions. We must move from interpretation to experience, narrative to performance, story to drama, and text to stage. We must become performing artists. If you want, move to Paris and perform without concern for being taken as a fool. As they say, those who jump off a bridge in Paris are in Seine. Cultivate your craziness, as Whitaker both taught and lived.

The history of therapy has mirrored the history of clients: both have grown accustomed to sitting still, contemplating their meaning. Both need to take a stand and act, not feigned action or any mindless action. What is required are whole bodies attending and minding the whole context, enabling natural action to be brought forth.

This is the Zen of therapy, the nonpurposive release of expression that cannot be premeditated or dictated by ideology, including an ideology of Zen or performance. This is the Tao of therapy, the embodiment of flow's mind, and the Wu Wei of spontaneous collaborative "we-nessed" (rather than witnessed) performance. This is the wisdom of the Kalahari Bushmen who dance themselves not into trance but into being fully awake.

As the grandfather cybernetician Warren McCulloch (1965) once advised, "Don't look at my finger, look where I am pointing" (p. 148). He as well as Gregory Bateson and contemporary ecofeminists are leading us to an "eco-logic; a way of thinking/knowing that is informed by metaphors of Nature's ways and is intentionally participatory" (Wegman, 2008, p. 1). Finally, all of this is the calling forth of the eternal mother who asks that we embrace, rather than understand, one another. Let us move to the stage that can hold everyone's voice, not through the voice of a standby solo narrator, but through the dancing chorus of the interconnected and differentiated familial plentitude.

Postscript

Beware the plotline with its deceptive allure that all is evolving in a forward progression. The line must be swallowed, like a string of spaghetti, entering the mouth of a never-though-always filled recursive dragon that cannot cease spinning a tale. Here we find every truth being a truth for only a moment and then quickly hardening into a lie if embalmed by theoretical fixing. Let us be grateful to all those courageous sojourners, whatever flag they may march under. They have sacrificed themselves, whether knowing it or not, so that a previously stuck knowing could be loosened and lightened. Their shining moment is placed on the altar of life and death, for what they say must also be loosened and released as it gets tired, worn, and eventually lifeless. Words, stories, and understandings are momentarily alive like moss, rabbits, and dragonflies—here for only a moment and then passing on. But in death, life is regenerated. Yesterday's ideas may spring to life as the biodiversity of mind dances unspeakable complexity. In this way, all our ancestors, women and men; cultural differences and unities; senses and nonsenses will have their seasons, in a roundabout way, in the turning that goes neither anywhere nor everywhere. Whatever can be said, narrated, known, evoked, or performed can never tie down and hold onto a single part or whole of the ever-changing *flux of complexus*.

CHAPTER 2

Three-Act Therapeutic Performances

The "Deer" Family

This is the story of a family in one of the poorest towns in America, a place located in the Louisiana side of the Mississippi Delta. There, an African American grandma takes care of five children, of whom the oldest boy was in trouble for fighting at high school. They live in a shanty house that is old and quite falling apart, with barely any trace of paint and with rotting wood. As is often the case when we enter shanties of families who are required by the law to see us, they turn off all the lights so you can barely see a thing inside the home. You quite literally sit in the dark. What follows is the story of the treatment of a family that was court ordered to see us for several months. This account enables a view of creative therapy that extended over multiple sessions.

Act I: A 9-Year-Old 15-Year-Old

In the dark, my colleagues Amanda McMullen, Ritchie Sheridan, Kristy Walters, and I did not address the oldest boy's fighting at school but immediately requested Grandma to imagine that she could magically snap her fingers and make one of the kids a role model for all the others. She said that would be Jack, her 9-year-old. Naturally, the older kids complained that Jack is her favorite.

As family therapists, we approached the contradiction and confusion that surround a 9-year-old who sometimes behaves (and is expected to act) older than he is. Jack needed to become comfortable with his actual age and make sure that the family was giving him an opportunity to grow older in a natural and positive way. I expressed my concern that a 9-year-old acting like a 15-year-old might be missing some of the skills, education,

39

and wisdom needed from the in-between years. I also reframed the acting out of the older brother's fighting in school as possibly contributing to trying to help teach Jack how to stand up for himself: "There is something to learn from rebelling and acting out." Grandma confirmed that the oldest boy was worried about Jack and that this all made sense.

In classic family therapy form, Grandma's concern and that of the older siblings became redirected toward Jack. I maintained a focus on how the family could help a 9-year-old successfully become a 15-year-old and, in turn, pave the way for becoming a successful adult who could survive on his own.

Act II: Surprising Love

At this point, at about 7 or 8 minutes into the session, I made a move for a new theme by saying, "Your grandchildren know your 'easy love,' but now it may be time for you to introduce some 'surprising love.' They need to see that you have some surprises they don't know about, and this is an important thing for you to introduce to the family."

"I'll try anything if it helps my babies," responded Grandma.

"Great! Why don't you hold a surprise birthday party for Jack with a cake holding 15 candles?"

Without missing a beat, Grandma replied, "That's easy. They love it when I bake a cake for them."

The following week, we were surprised to see that a light was turned on in the living room. For the first time, I could see the clients and the space where they lived. Sure enough, the family had held the surprise birthday party for Jack. Grandma said, "Everyone in the family has improved. I think this 'surprise love' has helped us."

As I sat listening to her, I looked around the room that I could now see and was surprised to see an enormous deer head mounted in the most prominent place in the living room. I also noticed that there was an aquarium in the kitchen with six fish.

Act III: The Creative Deer Family

This relationship began with a family who was court ordered to see us because of the eldest boy's fighting in school. We quickly moved to the middle act of Grandma introducing "surprising love." Next, we attempted to move forward to a completely out-of-the box frame.

I said to Grandma, "Would you continue with this surprising love? This is what I'd like for you to do. Get up early in the morning and start singing at the top of your lungs. If the kids ask you what you are doing, just say that you are excited about getting ready for a special deer party."

We discussed how she could bake a cake, get ice cream, and have a party for the deer. I gave her $10 to cover the cost of the party. She also agreed that, from then on, only the oldest family members (she and the oldest boy) could sit at the ends of the dining room table. Whereas Jack had sometimes tried to sit at one end, he was no longer to be allowed in that chair.

Grandma agreed to lead a family discussion, to be held during the serving of ice cream and cake, when they would choose a name for the deer. When they agreed upon a name, she was to walk them into the living room and stand under the deer head. At that moment, she would make this announcement: "If the deer could talk, it would say something to you that you have never heard before." Grandma was then to tell them part of a family secret.

I asked her if the family in fact had a secret, something the grandchildren had never heard. She said, "They do not know that their great-grandfather, my father, had been shot and killed." I pointed out what an amazing coincidence it was that their great-grandfather was like the deer: both had been shot and killed. She agreed that she would not say anything more than the statement that he had been shot and killed. She was to tell them that any more information could only be revealed at the next deer party.

In session three, not only were the lights on but the curtains were open, allowing the sun to shine in. With excitement, Grandma reported how she got up at 5 a.m. and sang "Silent Night" (keep in mind that it was early October) as loudly as she could. The kids were totally startled. She announced that they were going to have a deer party and, as expected, everyone laughed and teased her. Great excitement filled the home and, when the party was held, Grandma followed through with her promises. She changed the seating positions so that she and the oldest child were at the end chairs. The family named the deer, "Bill" and then went to the living room and spontaneously sang "Happy Birthday" to Bill.

The surprise event filled the family with a sense of curiosity, excitement, laughter, and teasing. The younger kids expressed strong feelings about their enjoyment of these activities, while the older kids teased the younger kids about it. Grandma said it was an experience they would never forget. Clearly, the family had moved their home context to being inside Act III, the reality of Bill the Deer overseeing the changes in their everyday life.

"What a special experience," I declared and then added, "Everyone will always be able to tell the story of the deer party. When the kids have their own kids and grandkids, the deer party story will still be alive and be a treasured part of your family's special history."

I asked which child would be most thrilled to receive a letter from the deer. Grandma said that would be the youngest granddaughter, Michelle, who was 10 years old. I told them to expect a letter in the mail from the deer. It would provide further instructions as to what the family could do next. That same day, we wrote "thank you" cards—one addressed to the youngest granddaughter and one for the family. The family card was immediately sent that day with an enclosed letter of instructions to Grandma. The other letter was mailed the next day. Here is what was written in the cards:

Deer Barnes Family,

Thank you very much for giving me a name. It helps me feel more real and more involved with each of you. Now that I have a name, you can expect to hear more from me.

Looking forward to all our future parties.

Love,

Bill

P.S. Please pass the enclosed note to Grandma

Inside the card were notes to Grandma and the youngest granddaughter:

Deer Grandma,

Thanks for my party and for getting to meet everyone. I am proud that you were able to say things to the family that have never been said before. I think you are finding a way to have more surprises and more unexpected kinds of love with each other.

This is what I'd like you all to do. First, it's OK for you to read this to the family. Tell them how I have a special view of the family from where I sit. I see how everyone works and plays together. But I am concerned that the fish may feel left out. They, too, need a home. Please give each fish one of your names. One fish needs to be called Grandma and the others need to have the kids' names. I think you will enjoy watching yourselves in a different way as you watch them.

This calls for another party. Here is some party money. Please make sure that the fish get surprised and don't forget to include me. I like to be looked at, talked to, and sung to and to watch you have fun together.

Love,

Bill Deer

Deer Michelle,

I want to say something special to you because I noticed the sparkle in your eyes and your special smile when you look at me. You are making me feel more real. Maybe I will see you in your dreams.

Love,

Bill

A week later, Grandma said that the family was experiencing no problems and that everyone was doing great. The kids told many other students at school about the deer and the neighborhood was talking about Bill. They named the fish and enjoyed watching them as a way of talking playfully about each other. They were surprised that the number of goldfish was the exact same number as the family living in the house. We discussed having the family put up a sign in the yard that says, "Bill lives here." Grandma then handed me a letter. The young granddaughter also had written a letter to Bill, the deer:

To: Bill the Deer
Bill Came in Our Life in 07

Hey Bill,
Today is your day. What you got to say about today is your day? We put a bandana around you. What you got to say about that?
Love,

Michelle

With Act III, we, the therapists, no longer had to work anymore. We just moved along with what came naturally from being inside a family with Bill the Deer. Because Halloween was coming up, we predicted that Bill would be sending another card with instructions. We sent this card that same day:

To My Deer Family,
It made me very happy to watch you give the fish their names. I also like to see you enjoying your parties. I'm excited about Halloween, aren't you? I wish I could go trick-or-treating with you. May I go? You could carry me around and let me say, "Trick or Treat."

I can't wait to see the sign that tells the world that I live here with you. I would like to ask a favor. During the day when you're away at school, I miss you. Could you put up your pictures next to me on the wall so I am not alone up here?

Grandma, I am planning a very big surprise that I will announce in a week. Please make sure that everyone gets ready for it. One more thing: please, please, please may I go trick-or-treating? If I can't go out, will the kids bring their friends to me? Here's money for you to buy the treats I will give away.
Love,

Bill

A week later, Grandma said, "We had a fun and funny Halloween. Other kids in the neighborhood came to our house just to see Bill. That deer has become the talk of the town. I even told our relatives about him."

We began preparing her for Thanksgiving because many relatives were coming over. I discussed how they could be introduced to Bill. I again suggested that they actually put up a sign in the yard that says, "Bill lives here." Grandma agreed to have the family make a sign and put it up in the front yard.

After Thanksgiving, Grandma said that all the relatives thoroughly enjoyed talking about Bill. "We put up some family photos around Bill, and we also put a sign underneath him that spells out his name. We also found an old radio and tape player at the junkyard and placed it near Bill. The grandkids now play music on it and dance in front of Bill. Oh, there's another thing I want to tell you. My relatives and I discussed my starting a catering business. I'm thinking of baking cakes and pies to sell."

I interrupted her to ask, "How will Bill be part of that business?"

"We already thought of that because I think we'll name the place 'Bill's Pies.'"

After that session, Bill wrote them a final letter:

Deer Barnes Family,
Thank you for making me a part of your family. I feel that I now can be called Bill Barnes. My initials are B. B.

Since I sit higher than anyone else in the house, looking over the family, you have made me feel like a king. Like another famous king from the Delta, I, too, am a special king who has music all around me. I would like to ask you for one more favor:

Please make a special new sign to place underneath me that says:

B. B. King

Thank you.

Love,

Your Very Own Special B. B. King

In a follow-up session, Grandma laughed and said they must have been changed by "all that surprising love. But, we've become concerned about the rundown appearance of the school and decided to organize a community group to go clean it up."

Keeping in mind that this is one of the poorest schools in America, I responded with a smile, "I'm sure that Bill is happy about this."

The case ended with the family having a relationship with the school that was the opposite of what they began with. In the beginning, they were assigned to treatment because of trouble and fighting at school. Now, they were organizing others to make the school a better-looking place. What can I say? There's nothing that B. B. King can't do in the Delta country!

In terms of three acts, we have:

A young 15-yr old ---------------→ Surprising love ----------→ The Creative Deer Family

Figure 2.1 Three-act therapeutic performance: The "Deer" Family.

Summary

At the start, we found ourselves working with a family that had a 9-year-old boy who was acting like he was 15 years old. Surprising him with a birthday party for a 15-year-old moved us to the middle act, where Grandma was in charge of introducing more "surprising love." This, in turn, led us toward a family life that creatively revolved around a deer head mounted high on the living room wall. Once they became the "Deer Family," one creative idea after another came to them, including taking care of the school that originally asked for therapy to take care of them.

This case teaches us that therapists should improvise—no different than the blues and jazz performers from our home state of Louisiana. Utilize whatever confronts you in a session and mix it up through a process that neither fully surrenders to the familiar or to the unknown. In this way, you will serve the muses of transformation, creativity, and improvisation.

The next evolution of therapy should involve more creativity and less piety surrounding the dictates of models, theories, and outcome studies. If so, we can expect all sorts of surprises: therapists who talk to deer heads or suck helium before they talk (yes, I have done that!), and a recognition that we are capable of performing many unexpected roles depending on the context that calls them forth.

The Queen of Shock

In this clinical session, a young mother reports that she has just moved back home, with her 20-month-old baby in tow. Her husband of 3 years believes that "she wants to be a mother more than she wants to be a wife." Their sex life is nonexistent, he reportedly has a drinking problem, and they are talking about a divorce.

In Act I, I listen not for more details about how the marriage has run aground but rather for the metaphors, or what I sometimes refer to as the "building blocks," that can be drawn upon to construct an alternative context in which the client can experience herself as resourceful. In this session, I identify and utilize several of these metaphors and then gradually weave them together to form a context for the client in which all the "facts" of her situation remain the same, but her contextual home for them has shifted entirely. She leaves this single session with not only a profound new sense of her own resources but also a concrete plan for reviving her marriage. Here, the work is summarized in three acts (*Note:* You can view the actual session as one of the DVDs in the *Brief Therapy Inside Out* video series hosted by Jon Carlson and Diane Kjos; the session is also described in Kottler and Carlson, 2004).

Act I: Melanie's Problems

Melanie (the client) began the therapeutic conversation by expressing concern over her husband's drinking and verbal abuse, her confusion about their marital impasse, and news that she had moved out of the home and was considering divorce. In addition, she said, "I have a baby at home and that's a lot of responsibility."

Rather than zoom in on her feelings of despair and confusion, I looked for any mention of a strength or resource in her life. In the first few moments, I observed that she was quite proud of being a mother. Sidestepping any further discussion of problems, symptoms, and psychopathology, I immediately started asking about her 20-month-old daughter. Melanie immediately lit up and offered, "She's such a talkative thing. She's just the joy of my life, so active, so alive. But it's so much work, too ..." She then worried about her husband again, tying him to the discussion of the baby: "I don't really trust leaving her with Sam because of his alcohol."

I deliberately ignored the choice to accentuate a problem theme and continued to focus on her ability to be a mother. I asked, "So you really want to be a good mother and do the right thing for your daughter?"

"Yeah. My husband thinks that I want to be a mother more than I want to be his wife, but I don't think that is true." Again, I chose not to address any pathological connotations but to continue underscoring the theme that she was a good mother. She was presenting both problems and resources. I focused on the latter so she could find a way to utilize her own strengths as we moved along. We had reached an opening to the next act.

Before proceeding, however, Melanie and I took a look at what she had to say about her husband's drinking. It was made clear that her husband still held a job and that what she characterized as problem drinking began after his brother died several years ago. His brother had been his best friend and "they shared everything together." Seeing this as an opportunity to softly reframe the drinking, I proposed, "So your husband has been conducting a 3-year wake for his brother?" She readily accepted this definition and proceeded to talk about her husband's family of origin which, she said, did not teach him how to talk to people or respect others. In other words, she was making the case that he did not receive good parenting. At this point, I had more than enough "resourceful building blocks" to walk us toward Act II.

Act II: Good Mother

The fulcrum for leveraging this case centered upon celebrating Melanie as a good mother and seeing how this provided both advantages and disadvantages. With respect to her impasse with her husband, I suggested that "maybe he's showing you how he doesn't want to be. He's hoping that you will respond in a way that can help him break out of his usual pattern." At that moment, she reacted by defining his drinking as imbedded within marital interaction rather than as his individual problem. She said: "He says that he is willing to quit drinking if the marriage works, but if the marriage does not work, then why bother to quit drinking?"

She went on to say that they only had one car and this required that she give him a ride to the train station every morning. Melanie subsequently reflected, "It occurs to me now that he's always had someone to care for him."

"So," I suggested to her, "that only confirms the idea that he considers you such a good mother that he wants you to take care of him as well as your daughter." We were clearly standing inside Act II, where the theme had shifted from pathological framings of Melanie's life to resourceful framings of her being a good mother. At that stage, we tottered back and forth on a variety of marital issues and her husband's drinking, but always doing so inside the middle scene of this being a celebratory talk about a good mother.

I began to tilt the seesaw farther away from problems and pathology by asking her to talk about an arena in her life outside her home that she felt good about. Melanie quickly responded, "I'm totally into my job," saying so with a huge smile of pride. She went on to say that she worked with handicapped people. Without missing a beat, I responded as a cheerleader, "So everything in your life seems to be connected to you being such a wonderful mother—to your daughter, to your patients, and even to your husband." No matter what she said, I was now able to use it as a building block to further establish our presence in the middle act, where all discourse congruently reinforced and applauded the reality of her being a good mother.

When the middle act was sufficiently stabilized, I was able to articulate the dilemma that her resourceful presence in life brings: "But this creates a dilemma for you. On the one hand, you say that Sam's behavior brings you a lot of pain that is making you crazy, but his childish actions are also a way to let you know what a great mother you are; otherwise he wouldn't act so much like a child. It's as if he's testing how good a mother you can be."

Melanie nodded her head in agreement. "Yeah, I see it now that you point it out like that. It's like I'm trying to help him grow up."

Continuing to stabilize this fulcrum for transformation, I continued: "This is certainly a crazy way to compliment your mothering skills." I showed her how all the complaints she had about her husband could be seen as crazy compliments and requests that actually honored her good mothering skills. This notion was accepted in a way that would have been rejected in the beginning of the session. Why? Because it was articulated in Act II—that is, it was held inside a different house of meaning where it made new sense. We were now in the contextual home of a "good mother" rather than a "sick marriage."

In less than 20 minutes, I was no longer hearing complaints about Melanie's life but rather reports about her skills. At that point, I could positively challenge her and see if the fulcrum could push us farther toward a creative transformation.

"I'm wondering if you can recall a time, even just 5 minutes in the course of a week, when you don't feel like a mother to Sam, when you …"

"No," Melanie quickly interrupted. "There really isn't."

"How about 1 minute then? Can you think of a single moment when you relate as wife and husband?" I wouldn't back off from gently showing and teasing how her reality is entirely whirling around the distinction of her being a good mother.

"I don't honestly know," she said while shaking her head. "When we first met, we were soul mates. I still sometimes feel like I connect with him. We can almost think for each other at times. It sounds weird, as much as I could strangle him at times, I also feel like I know what he is going to say before he says it." She momentarily pauses and adds, "And we used to go for long walks and talk all the time."

At that point, we had an opening to move Melanie out of the middle act toward a final act, where she could bring back the possibility of being a wife and not just a mother. "If the two of you would go for a walk right now, where would you go?" For several moments, she found one reason after another for why it wouldn't work out for them to have a walk together any time soon. So I reminded her of all the things she had said (in Act II) regarding her skills—in her own words, she had "good analytical skills in solving problems" and she was "good at figuring out how to get people to do things." Equipped with her own attributions of skills, I then asked again, "I bet you have the skill to figure out how to get Sam to go on a walk."

"Well, probably if I tried really hard, I could."

I was near an exit from the middle act. I immediately asked her to continue moving forward and I did so with a slight tease and an air of conspiracy: "Melanie, I'd like you to turn on your imagination for a minute. I want you to picture a way that you could ask Sam to take a walk that would completely shock him, that would knock him out of his socks."

"Hey," she responded playfully. "I'm a mother, remember?"

Yes, she was, without a doubt, rooted to the middle act but playfully considering how to move on. We were enjoying one another's company and we laughed together as I continued, "Let's just say that you wanted to do something that was weird, mysterious, even a little crazy, something that would capture his attention. This could be an idea, a gesture, a phrase, something, anything."

"A gesture maybe."

"A gesture. That's good!"

Act III: Queen of Shock

Then there was a pause. The momentary silence was pregnant with possibility. I felt like I was helping her give birth to her inner imagining by further encouragement to push a little harder. "You know, the really nice thing about pretending is that you can just make something up. I'm not asking you what would work. I'm just asking you what you think might shock him."

As if giving another reason for why she couldn't do it, she replied, "I've shocked him so many times before with other things."

"Oh?" I riveted my attention to this opening, this possible exit sign. "So you're really good at shocking people?"

"Yeah, I'm very good. My mother says I'm the queen of shock."

With enthusiasm and excitement, I responded, "Really? You're telling me you are the queen of shock and you can't come up with something that will get Sam to take a walk with you. I think you're holding back."

Melanie giggled. She said that she must have watched too many Barney movies with her daughter and that may have made her out of practice.

Act I was about a problem husband, a problem marriage, among other problems. It may be called "Melanie's life is a problem." Act II moved us into the theme, context, conversational home, or reality of "Melanie is a good mother." Now, we are in Act III, led there by a road sign that points to "Melanie, the queen of shock."

"Okay, how have you shocked Sam in the past? I mean you are the queen of shock and all."

She described how she used to send Sam secret beeper messages. If she wanted to tell him that she loved him, she'd do it by sending a sequence of numbers that was a code he had to break. "It probably took him half an hour, but he really liked it."

"So you really do have a special talent for shocking Sam. And it's proven to be wonderful for your relationship."

She shrugged and indicated a modest agreement.

"If you were going to invite your husband to go on a walk and do it by sending him a coded message on his beeper, how would you do that?"

"I'd just write it out with a pen first and then code it in." Clearly, she was thinking about this as the queen of shock, someone who is an expert at such a task.

"We've hit some interesting territory. That is, we've found a skill that you held back from telling me." I continued to playfully tease her, "I wonder if you've ever done anything that even shocked yourself?"

"Not that I can think of."

I asked her again to imagine something so shocking that it would even shock her. I was asking her to access her own creative expression.

"I just don't as much anymore," Melanie admitted. "Sam tells me this too. I just don't joke around the way I used to."

I saw this is as an important transformative moment. The time was right for me to ask: "You know what I think you need to do?"

"What's that?"

"You need to announce loudly and clearly that the shock is back in your life. You need to say to yourself, 'I'm bringing back the shock. The shock is back!'"

"The shock is back," Melanie voiced with a giggle. At that moment, transformation clearly and unambiguously could be seen. She was no longer talking about her problems but about her resources. She came to the room defeated and now was a queen, the queen of shock. We continued stabilizing this shift in her present life by talking about all the ways she had been shocking in the past—playing tricks, being childlike, laughing, teasing, and feeling more alive with unpredictability. Melanie mentioned that she used to hide things from Sam, making him hunt through a "maze" to find them. She also set up a treasure hunt that he had to go on to find a gift. After each example of how she was the queen of shock, surprise, unpredictability, and creativity, I cheered and applauded her unique talent and way of being in the world.

"It sounds to me like you are willing to announce to the world that the shock is back."

"Yes, I guess that's true."

"You can be a good mother and yet still be a good, a great shocker." I went on to say, "One does not exclude the other."

"Yes," she agreed.

"What this relationship needs most is some shock. And you're just the person to do this."

Now that we had moved from a problem marriage to a resourceful mother and a master shocker, it was time to discuss bringing forth a creative marriage. I widened the discussion to include how she could recruit Sam into a shock conspiracy.

"Have the two of you ever shocked anyone together?"

"Hmmm," she pondered with fascination over the possibility of doing so. "I don't think so."

"Well, now that the shock is back and you intend to beep him to take a walk together … while you are walking, you can plot together how you might surprise someone. Who might that be?"

Melanie, now physically looking like an entirely different person than she did at the beginning of the session, smiled and enjoyed pondering the possibilities for her new future—being part of a husband-and-wife shock team. She glowed and nominated her sister-in-law as the first target for their shock plot. She was the person who was most likely to enjoy a practical joke. Melanie, now a smiling, laughing, animated, and spunky mother and wife, was celebrating and rejoicing her creative reentry into life.

As the session moved toward the end, I asked again: "So you promise me that the shock is back in your life? I want you to promise me that the shock is back into your life for at least 1 week."

"A whole week? I can promise you that."

In closing the session, I reminded Melanie that life always faces many challenges and adversities. That was a given. What I wanted her to remember was that she must draw upon the creative gifts and resources she has within her. "You have a gift and you must use that as best you can."

Toward the end, we reached to shake each other's hands. I momentarily hesitated and said, "What if you have one of those trick buzzers in your hand and it gives me a shock?" We burst out laughing, shook hands, and celebrated the fact that the creative queen of shock was back in action.

Summary

In this session, a whole therapy was performed involving three acts:

Problems ------------→ Good Mother -----------→ Queen of Shock (and

shocking couple)

Figure 2.2 Three-act therapeutic performance: Queen of Shock.

In the beginning act, I did everything to avoid the temptation of being curious about her problems. I said no to all invitations to be an archaeologist of pathology, a fixer of present problems, or a hunter for new solutions. Instead, I looked for an exit and found a conversational road that led us to the middle act, where I could celebrate and talk with a good mother. There, everything that had been previously discussed could now be experienced as evidence (metaphors or building blocks) for the alternative reality of "this is all about her being a good mother."

When this middle fulcrum was sufficiently stretched out, there was enough leverage to tilt it so that movement forward would tumble forth with little effort. This led us down the trail of rediscovering her strength as a queen of shock. As we entered her imaginative kingdom, we awakened a humorous, teasing, inventive, creative queen who has all the gifts she needs to transform her own life, relationships, and situation. Whatever happened to her in the future, this session accessed her own resources and reawakened her own creative abilities to handle the challenges that came her way. Rather than "fix a problem" or "find a solution," her natural abilities were brought forth so she could more creatively navigate her own life.

The Light Man

The next example of a three-act performance of creative therapy took place inside the walls of a state prison. In this case, we see that there are ways to break out of any box, even one that has barbed wire fences and armed guards.

Act I: Unfair, Rotten Deal

Greg was a 17-year-old who had been incarcerated for a year after being convicted of a "sexual battery" offense. He had helped a 35-year-old man enter the window of a house, where the older man had sex with a minor in a "sex-for-marijuana exchange." Greg felt he had been set up and did not deserve punishment for something he did not know was going to happen. In prison, he seldom read or engaged in much activity, was truant in the school, and went back and forth between feeling angry and depressed about how he had been given an "unfair, rotten deal."

In our first session, my colleague, Scott Shelby, and I began by asking Greg what had happened in the last months that made him proud of himself. He said he had learned to do some electrical work and was helping perform minor electrical repairs around the institution. He learned these skills in a recent trade program. In this discussion, he expressed a glimmer of hope for his future. Greg went on and on about how the maintenance staff was asking him to help them do electrical work. He tried to remember every light fixture he had installed.

We allowed him to express his enthusiasm. He was talking himself right out of Act I and into the next scene, where we could work with him in a more resourceful manner. In other words, following Greg, we exited from Act I, the "problems of prison," to entering the context of "an electrician in training."

Act II: An Electrician-in-Training

We continued exploring the theme of Greg making a future living as an electrician. We brainstormed about possible names and slogans for his potential business. He talked about the various things he was learning in the electrical trade class and how the teacher sometimes asked him to help out with some of his chores around the yard.

"Did you take that course on printing?"

"Yes, sir," Greg replied. "And we learned to make our own cards."

"Excellent! Maybe you should make yourself some cards for your future business. What would your card say? What will you call yourself?"

Greg could hardly contain himself, "That's an awesome idea!" He paused as he thought about all of this. "I got it. I will be the Light Man. I want to become the Light Man. Call me and I'll install a light."

Utilizing his suggestion, we began calling Greg the "Light Man" and he referred to himself that way. This was our entrance to the next act.

Act III: The Light Man

Greg wanted to come up with a slogan for his business card. We discussed several ideas and he decided he would print this card that was written out for us in the session:

"CALL THE LIGHT MAN … I BRING THE LIGHT!!!"

He discussed how he would like to bring the light to others. The first thing he wanted to do when released from prison was help others, beginning with his mother, with electrical services that would bring more light into their homes. To our surprise, he expressed an interest in wanting to draw lightbulbs all over the prison, placing them on pictures, drawers, shelves, walls, and ceilings. He was proposing that the image and metaphor of a light could be added to practically anything in his surrounding.

In this first session, Greg also started to sing a song about his new identity. He confessed that he was not a great songwriter but knew someone at the prison who was a musician. He decided to ask this other inmate to write the "Light Man Song." He excitingly declared that he wanted to run out of the session and tell everyone his new name. He then shouted out to us, "I bring the light of joy!"

Hearing his enthusiasm, I leaned over and said, "I'd like for you to try a creative experiment tonight. Draw a lightbulb on a piece of paper, and when you've done that, write the name 'Light Man' inside the drawn bulb. Can you do this?"

"Yes, sir. That's easy. I like to draw."

"Tonight, before you go to sleep, place this lightbulb under your pillow and sleep on it."

"Can I color it yellow?" he asked.

"Absolutely. Do just that and feel free to make any other interesting modifications," I replied.

The session ended with our agreement that he would sleep on his lightbulb that evening. Later that day, as we were walking outside the yard, the Light Man opened the door of his dormitory and enthusiastically waved while he proudly shouted at the height of his lungs, "Hey! I'm the Light Man!"

The following week, we found that the Light Man carried out his assignments and made a business card with two lightbulbs drawn on the bottom corners. The "Light Man Song" was being written in the musical genre of hip-hop. He called his mom that week and informed her that he was now the Light Man. Most importantly, he drew the image of a yellow lightbulb and placed it under his pillow. To his amazement, Greg had a visionary dream:

> I dreamed I was in an aluminum fishing boat with an old wise man. We were on a swamp filled with alligators. While we were in the boat, I held a light for the old man to see the gators. The old man had gray hair and he had a friendly smile. Oh yeah, he was also wearing some alligator-skin boots. The man told me that he was gonna teach me some things. First thing he taught me was how to shoot an alligator with a crossbow. He also showed me his right hand and I could see that his right index finger was missing. He told me that a gator bit it off when he was young, but that today he doesn't have any fear of them getting him. We talked about a lot of things. I was really surprised when he told me that he'd be back. He said, "I'll be back." Then I woke up.

After hearing this dream report, we encouraged the Light Man to draw an outline of an alligator under his bed.

"The alligator you draw under the bed should be the same size as the bed. Now, after you draw it, I want you to know that, from this night on, your bed is never to be called a 'bed.' From now on, you must call your bed a 'boat.'"

We also encouraged him to draw the light he saw in his dream and keep it by his side at night. We encouraged him to think about the old man's last words to him: "I'll be back." We told him that he could ask the wise man anything he wanted, including questions about the meaning of life. We explained how such a dream was called a "vision" in other cultures and that it was regarded as a *very important experience*.

Everything we did in the second session was nothing more than help him elaborate and legitimate his presence in a creative context. Now he was more than a prisoner, or an electrician-in-training, or a light man. He was becoming a visionary, a dreamer in an imaginary boat who went outside the prison walls every night in order to learn some secrets to life.

In the third session, Light Man reported "good dreams" that he enjoyed very much. He had drawn the picture of an alligator under his boat (bed) and proudly expressed that he told several people that he goes to his boat at bedtime. He slept with his drawn pictures underneath his pillow and asked the old man, who he now called the "Alligator Man," to visit him.

"I've thought a lot about all of this, and I believe that these dream messages are coming from God," he said.

I replied, "Then it would make sense for you to place your drawings inside your Bible and sleep with your head on your Bible. I think that might be a good thing."

"That is a good thing. I bet that will supercharge everything."

I added, "In addition, you should assume that the Alligator Man always visits you every night, even when you don't remember him. I think you should say, 'Thanks for seeing me' instead of 'Will you come see me?' before you fall asleep. I reckon that the Alligator Man does come to you every night and that he comes in different ways, sometimes you see him and at other times he is invisible. Whether you catch him in a dream is sorta like getting lucky when you go fishing. When you go fishing, the fish are all around you; whether you catch them or not, or see them or not, they are out there."

We discussed how the Alligator Man had cut the Light Man's prison sentence in half. He was only in prison during the daytime because during the evening he went outside the prison walls and explored the world with his teacher, the Alligator Man.

Greg mentioned that when he was a young boy, he once went out hunting alligators and was bitten on his right pinky finger. He did not lose the finger, but it now surprised him because the Alligator Man he saw in his dream was also missing that finger. We began planting the seeds for the idea that the Alligator Man could be many different people, like different pieces of a jigsaw puzzle, that have been put together as an amazing guide to help him on his life journeys. We did not say that the Alligator Man might be who he was becoming.

In our subsequent sessions, Greg continued to tell us about the special dreams that connected him with the Alligator Man. He described a dream where he assumed the role of a teacher who helped his friend not be afraid of the alligators in the water. Greg described how he faced the alligators fearlessly while in his boat and that he reassured his friend that things were all right. Greg also stated that he had other dreams that included boats and recreation where he used kneeboards and inner tubes to have fun. He was learning that after he cleaned up the troubling waters—clearing it of alligators—he would be able to have some fun. Greg started seeing how he was taking the place of his teacher in his dreams and he was teaching others. This made him very excited.

He learned that some of the other prisoners and guards were the alligators in his present prison life and that he needed to stay away from getting in the water with them because "once you get in the water with alligators, they will bring you under and possibly kill you." Greg wanted to tell off some of the staff, but he knew that would be like putting your foot in the water for the alligators to chomp on. We discussed how he was safe in his boat with his light shining into the darkness.

Greg, the Light Man, became a model prisoner, and was released early for his exceptional performance as an electrician. I told him that I had played some blues on the piano while telling his story to an audience at an international conference. He was thrilled and requested that we make a recording of my performance for his mom. A short film clip was given to Greg as a final parting gift.

A few days before being released from prison, the Alligator Man came to Greg in a dream and said: "I have something to tell you. I'm going to pass my job on to you. Now you need to be the Alligator Man. I am no longer needed because you now have what it takes to stay out of trouble."

"I told him that I thought he was right and that I was willing to take on this responsibility," said Greg, who was now both the Light Man and the Alligator Man.

He added another comment: "You know, I figured that if I am talked about on a film that has been shown across the world, then I must have what it takes to answer things."

Greg's last message to us was, "I hope the Alligator Man knows how much I appreciate what I learned from him while I was in prison. God is my light and I plan to bring the light to others!"

Summary

This therapy, with numerous performances over the months, quickly moved from the opening and middle acts to a creative and resourceful ongoing final stage:

Prison ------------→ Electrician-in-training -----------→ Visionary Light Man

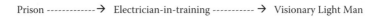

Figure 2.3 Three-act therapeutic performance: The Light Man.

Once Greg entered a context that called forth his creative imagination, we simply joined his natural movement. We handed our role over to his dream teacher, the Alligator Man. As that elder taught Greg how to stay out of trouble and clear the waters from harm, he learned how to help others do the same. His creative life was deeply accessed through the visionary boat rides taken on a bed in prison. To our surprise, months after Greg was released, we discovered another prisoner who was a client of ours was receiving letters of advice from the Light Man. Greg now shines a light for other prisoners wanting to know how to get on with their lives in a creative manner.

The Medicine Man Who Never Had a Vision

This session (also described in Kottler & Carlson, 2003) has practically no conversation, but it remains a three-act whole therapy. Nate was an ex-con who had served hard time at Sing Sing. Burly and muscular, with scars on his neck and face, he looked the part. He also had a long braid of hair that went all the way down to the bottom of his back. He was called a medicine man by his tribe in South Dakota and was a leader among his people. He never claimed to own the role of medicine man. Instead, he chose to help other medicine people conduct their ceremonies.

Act I: "Help Me Have a Vision"

Nate came to my home and without wasting any words confessed: "I am ashamed to tell you why I have come because I would not want my people to know. I have never had a vision … I have never personally had a vision. I don't know why. I have talked to several elders. They have put me out on the hill to fast and pray. I have done all these things and have done them sincerely, but I have never had a vision. I want you to help me have a vision."

Act II: Ceremony

"Nate," I replied, "let's go downstairs and pray in a traditional way." With this invitation, I avoided therapy altogether and shifted the context to his culturally prescribed way of transforming personal issues, something I was familiar with because of my personal involvement with his culture. In the setting of a dark room, with traditional heartfelt drumming and singing, prayers were simply expressed. We did not know what to expect. His concern was sincerely handed over to the unspeakable, unknowable acknowledgment of mystery. Such a setting invited the deepest part of the mind to express itself with no need for conversational exchange. This we did in the dark with only song, drumming, and prayer. There was no discussion about anything.

Act III: Visionary Experiences

In this ecstatic ceremonial encounter, I had a vision. I saw Nate go into the woods and gather 12 branches, each no longer than a foot in length. One branch was cut from a tree facing south, another from one facing north, and the others from points in between. I told him he should consider doing what I saw in my mind's eye. "Mark each of these branches carefully so you can tell which direction it came from. As you know, you should ask permission to be led to each branch. Then take the branches home and place them underneath your bed in a circle with a traditional tobacco offering. Each branch should face its original direction, with the south facing south, and so on."

Without any further discussion, Nate understood, thanked me, and went home. He gathered his branches and prepared to sleep over them.

Within a week, I received a call from Nate. "Man, I'm just blown away. I don't know what you did or what it was about, but I went out and did what you said. I found the branches … and put them under the bed like you told me. And then I went to sleep. I had this dream. But when I woke up, I was still in the dream. I was flying out of my body and flew back to a time when I was a little boy, and I looked at my dad and my family and then I realized that I was flying and was wide awake and it freaked me out. Then I woke up."

"Well I guess you had yourself a vision."

"Brad, man, that was too much. I'm not sure I want to put those branches under my bed again."

I teased him about maybe he put too much tobacco out as an offering. We both laughed.

"I think I know what I must do now," Nate declared. I remained silent with no need to explore what that might be. We said goodbye to one another on the phone.

A few months later, I found out that Nate found a way to work out some issues with his family of origin. That learning helped him become a spiritual counselor to adolescents in trouble with addictions.

Summary

This three-act therapy moved us straight through these themes:

"I need a vision" -------------→ Ceremony -----------→ Visionary Experience

---

Figure 2.4 Three-act therapeutic performance: The Medicine Man Needing a Vision.

In this session, there was no therapy talk, only a simple statement of what Nate wanted, followed by the utilization of the healing practice that is part of his cultural tradition. This led us to visionary guidance that helped fertilize, seed, and germinate his vision. However, there is more to the story because both our visions seemed to trigger another vision from someone who didn't even attend the session.

Nate's wife, Betty, and his friends, Ken and his wife, Joan, later came back with Nate to see me. His friend said something I will never forget. He began by saying that Nate told him about those branches.

"After I heard about putting the sticks under the bed," Ken went on to say, "I had the weirdest experience of my life. I don't know how to make sense of it. I had a dream where I was sitting in the lodge [of a traditional ceremony]. There were medicine people and a big fire and hides covering the wigwam made of branches. I saw my grandfather. He came over to me carrying red paint in a seashell. He took his right finger and dipped it into the paint, and he drew a stripe right down my forehead. It freaked me out because when I turned around, I saw that all the men in there had animal heads. I don't know if they were masks, but they looked like real animals."

Ken paused and took a deep breath. He looked at his wife, and Joan nodded, urging him to continue with his story.

"I woke up real startled. I was sweating and screaming, and I woke up my wife. She jumped out of bed and turned on the light. Then she screamed. What my wife had seen was that I had a red stripe on my forehead."

This man was the head of a mental health agency and was not prone to make up a story like this. The experience had shocked him and his wife. I found out that Ken had been struggling with whether to go back to his traditional ways and join a medicine lodge, but his Catholic upbringing filled him with doubt. Now he felt the presence of a mystery that could neither be explained nor denied.

Not only had Ken not sought help from me but the task from another person's session somehow contributed to a transformative experience in his own life. One ceremonial vision led to a home vision and then to another visionary experience of the client's friend. The ways in which creative expression enters therapy will always be a mystery. You might say that all of therapy is held in the dark. Only in an empty dark space, that which holds our own emptiness, can we possibly have a glimpse of a new light, of a line pointing the way to the mysteries of creative transformation.

CHAPTER 3

In-and-Out-of-the-Box Musings

Typically, one of the beginning chapters of a traditional clinical book sets forth a logically consistent theoretical framework that spells out the ideas that underlie the presented clinical work. I am breaking that tradition because I believe that the hegemony of theory over practice is problematic. Here, my intent is to provide some thoughts and notions that aim to jar the therapist's comfortable relationship with theory, particularly any overly tight hold of a model, school, or orientation. I do this as a means of helping release the natural flow of creative action.

Academies of the performing arts sometimes present exercises to help loosen up the actors, ranging from mumbling nonsense sounds to whistling like a bird or roaring like a lion. Consider the following musings to be "thought exercises," not to be taken too seriously, but as something to help stretch you and get you thinking about different ways of thinking about the performance of creative therapy. Playfully consider them as a means of helping you stimulate and stir your creative juices.

When we stand inside the theatre of creative transformation, the last thing I want to offer is a monologue on why we need another model, meta-model, or antimodel of therapy. While on the performance stage, I don't wish to kill therapy's creative life in order to pin it on a conceptual board, making it like another dead butterfly in a naturalist's collection. At the same time, I recognize that there are some notions that are useful in helping set us free from restrictive models (and all models are restrictive!).

I therefore offer an assortment of musings that pragmatically aim to liberate us from any framework that imprisons our natural ability to bring forth creative therapy. These notions are not intended to be steps toward any theory. I have scrambled their order so that they have no intended logical connections, necessarily consistent relations, or patterned sequences. Any ordered cognitive schema or hint of a theory is to be regarded as the construction of the reader. I take no responsibility for the understandings or misunderstandings provoked by the following text, as no rigid understandings are intended. In no particular order:

Pay the most attention to what is transformative: All experiences hold the potential to creatively contribute to one's growth. This refers to ideas, beliefs, understandings, attitudes, events, conduct, habits, dreams, memories, fantasies, sense, nonsense, feelings, body sensations, movements, social interactions, community involvement, cultural immersion, and any other form of experience. Creative therapy focuses on bringing out the transformative nature of *whatever* a client presents and performs. It utilizes what is on the table, using it as grist for the transformative mill. Strictly speaking, creative therapy does not treat; it helps transform. It may help turn sour lemons into sweet lemonade as well as turn them into a slippery banana peel. At its best, it does not appear as therapy. It tries to become an unpredictable, spontaneous *theatre of creative transformation,* with performances that aim to awaken both the client and therapist, to move them out of the game of trivialized and clichéd psycho-babbling exchange while stepping onto the broader stage of living the everyday with more creative gusto.

Pay the least attention to pathology: Creative therapy pays little attention to problems that seem to be in search of ways of becoming more pathologized. This means no use of any diagnostic nomenclature. Or if a label is already present, do whatever is possible to deflate its authority and importance. Creative therapy is also careful to avoid creating detailed behavioral descriptions of any problem or mythologizing them through medical, cultural, postmodern, narrative, alchemical, Grecian, or astrological metaphors. The last thing a client needs is more problem talk (or its inseparable Siamese twin: solution talk).

Clients almost always enter therapy after many dress rehearsals with others in their social milieu. They have also privately rehearsed. Paid-for therapy differs in that it escalates the game to having more authority and legitimacy. The latter is crassly accomplished by medical and psychological testing. It is also perpetuated through the carefully crafted pontification of grand interpretations—whether inspired by Sigmund Freud, Michel Foucault, or Simone de Beauvoir. The art of therapeutic healing requires paradoxically delegitimizing "clinical staging" while holding clients' hands (or kicking their posteriors) and leading them to a more transformative way of being in the world.

De-emphasize the roles of "therapist" and "client": The creative therapist is not interested in maintaining the traditional roles that make up the cast of a stereotyped therapy performance. Instead, each participant is regarded as being caught in a similar interactive relational dilemma: how to wake up the other person so that the session becomes creatively alive. The challenge is to turn these individuals not into dancers, but into a dance. The dance of therapy, when it is flowing, is nothing less than what a systemically oriented thinker would call the "mind of therapy."

What must the client say to transform a therapist into not being a therapist and what must the therapist say to help change the client into being a nonclient? Would it help for a professional improvisational actor to enter the room and move things along? Probably, as long as the actor wasn't inducted into the performance genre of "therapy." The actor must think he or she is on a theatrical stage and that the audience wants to see something go beyond the habitual expectations associated with therapeutic discourse. We must look for a creative turn that leads the clients and therapists to being transformed into more interesting and less predictable human beings.

The creative therapist wants the least amount of clinical information: The last thing the creative therapist wants is more talk that contributes to legitimizing that he or she should be doing therapy. Whether it is about the past or present, problems or solutions, individual or family explanations, archetypal dramas, biochemical assays, narrative or postnarrative editorials, failed or successful solutions, diagnoses or counter-diagnoses, this kind of talk buries us in the situation we are trying to escape. Therapy does not have to be a "no exit." A client's efforts not to talk about therapy may not be "resistance" but rather an invitation to find a way out of the clinical quagmire. Talking about cooking, whistling, fishing, or gardening is more likely going to take you somewhere more resourceful than redundant clinical chatter.

Remember, unless your client is willing to pay an enormous sum of money for a lot of sessions, teaching him or her to talk a new abstract language in order to voice a therapy script probably isn't going to work out. Therapies that require indoctrination in abstract talk, whether it includes archetypes, transference, astrological signs, family constellations, or cultural criticism, don't necessarily contribute to a focus on *what* the client presents as a metaphor to work with or *how* the client talks about it as showing a direction for how to work.

The goal of Act I is finding a clue for how to exit stuck talk, action, and experience and find a more inspiring subject matter. In this regard, we are following Maslow's early advice to the field to stop focusing on pathology and sickness and turn ourselves to the topics that concern growth and the actualization of being more fully human. Our clients bring road signs that point to their desired forward movement. It is our responsibility to follow them. If your client is the ghost of Carl Jung or a Jungian, then talk Jungian abstraction, but only if the client suggests that kind of discourse. If your client plays violin for a symphony, you may find yourself discussing how to help him tune his other strings.

Stop and ponder whether you want to ask a client, "What is the problem?": If asked this question, clients will typically respond that something is the problem. They may even say that they don't know what the problem is and that *that* is part of the problem. After this inquiry, ask again, "I mean, what is the real problem?" They will most likely say something different. Ask one more time, "As I speak to your deepest unconscious mind, I ask it, 'What is the deepest problem you are carrying?'" Still another answer will be set forth. Be careful what you ask for. If you ask for it, it will come. The client's world and yours is not discovered. It is co-invented.

Stop and ponder whether you want to ask about outcome: Ask a client, "What has happened to you since we started working together?" The client will say something. Now ask the question again, "I mean, what really happened?" You know the next question. Which outcome report is the most real? Be careful what you ask. Invariably, the client's world, including its reported outcomes, is co-created. Your question and the researcher's inquiry about outcome is another therapy, another intervention, requiring another outcome study ad infinitum.

Be careful when you ask other therapists to "explain" what they did: They might answer like a famous jazz pianist once did at the end of an amazing concert performance. His first response involved pointing out all the technical skills and harmonic progressions he made at the piano. The reporter persisted, "What's behind that? What really moved you tonight?" The jazz great then articulated a theory of music and how it was related to what he was feeling that evening. The reporter did not stop her inquiry. "What is the deepest understanding that you hold about music that was responsible for what we experienced at the concert?" This genius musician proceeded to give a nonstop, nonpersuasive lecture on scientology.

I submit that the same thing often happens when we ask performers of therapy to tell us what is really going on with their therapy. They eventually respond with metaphysical-like theories that, as true as they may be for the therapist, don't help creatively awaken any listener. Leading therapists do this in their writing, at conferences while in dialogue with colleagues, and sometimes with their clients. Be careful what you ask and how you answer.

It's not about any particular thing that is said, it's about the context that holds it or the frame that frames it: Frank Zappa (in Barron et al., 1997) declared that "the most important thing in art is **The Frame**" (p. 196). He meant his statement figuratively for all the arts. As he put it, in the case of painting, "You have to put a 'box' around it because otherwise, what is that shit on the wall?" Without talk being framed as "therapy," it's simply conversation. Asking, "How are you today?" is harmless unless asked by a psychiatrist in a mental hospital, who doesn't smile but frowns and stares and has a tone that implies that something must be wrong (with you).

Now consider framing that frame within a meta-frame of humor. Now you have a possible comedy act. Imagine Robin Williams enacting this same mental hospital scene, but this time it's part of a stand-up comedy routine. In this new and wider frame, our comedian can say unexpected things to the patient that a patient might never imagine hearing in a mental hospital. The patient might answer the doctor's inquiry by responding, "I don't know how I am. Let's ask me." If the line could be transplanted to the hospital and delivered with a wink (and some audience applause), perhaps the psychiatrist could momentarily rise above the double-binding nature of the interaction and experience its absurdity. That would provide an opening that helps each escape their self-verifying dilemma.

This is the way out of nontherapeutic therapeutic encounters. Place all potentially toxic frames or any of their content inside new frames that bring forth more transformative possibilities, such as humor, exploration, virtuosity, curiosity, play, music, poetry, cooking, carpentry, or any other creative *road to awes*. All you need is a heart, a brain, and courage. You were born with the first two. Now be courageous.

I'll say it again because it can't be said enough: it's about the frame, not the contents inside the frame: You can sometimes do the wrong thing in the right context and find that it's transformative. And you can do the right thing in the wrong context and discover that it doesn't change a thing. Contexts (or framings) rule, not the elements inside them. Systems of relational interactions organize what takes place, not their individual members.

Everyone knows that clients should stop doing those damn nonresourceful behaviors. But they persist. Why? They continue recycling a noncreative stuck habit because it fits inside the frame they are living in. Similarly, this points to why some things work that shouldn't work. A communication is effective when the framing helps it be transformative. For example, consider asking a family, "Who would be the most pleased if Dad were swallowed by a whale?" As wrong as this might seem to actually ask, when delivered inside the frame of an absurd tease, it may serve to help open an entangled knotty situation.

I invite you to make a list of three supposedly "safe" therapy lines, like "I hear what you are saying," "How does that feel?" or "What would your mother say?" Now imagine framings where the utterance of these lines in a session results in trouble. When that is completed, do the opposite. Come up with three unimaginable therapy lines, things you'd never imagine any therapist saying in a session. These utterances, without appropriate context, may sound too wild, dangerous, uncomfortable, or irrational. Here's one I have used as an exercise for therapists: "Most people write suicide notes. Why don't you consider writing a suicide comic book, or a suicide musical, or a suicide miniseries?"

Now put on your creative theatre hat and imagine a playwright inventing a framing where this kind of discourse would be transformative and therapeutic. Consider doing this assignment in the role of an imaginary professional assassin. Your task is to take out any clichéd therapy lines that are assumed to be therapeutic regardless of the context within which they are mumbled. Aim and fire away!

Actually, it's not about the frames inside therapy; it's about holding therapy inside a more encompassing frame: In spite of what I have said about leaving therapy, know that it is impossible to leave it. You are working inside a context of therapy. However, you can place that frame within a wider frame and then the tension between "therapy" and "not therapy" may jump-start some creative flow. The more you posture yourself as both a therapist and as a nontherapist, the more creative tension a session will generate.

This also applies to framing yourself simultaneously as a schooled therapist and an unschooled therapist, a theoretically oriented clinician and a nontheoretically oriented one. You must hold one frame within another, and better yet, have them oscillate back and forth, in and out of one another, like Chinese boxes. What was inside is now outside, now it's inside again, and on and on it goes. How you do this in a unique way for each client is the art of being a practitioner of creative transformative theatre.

Do the least: Use the least amount of time needed to launch a session and then let it go. Also aim to have each session be the end of a whole therapy. If the client shows up the next week, consider him or her to be a new client—the client you saw once before plus what happened to him or her since you last worked together. Limit yourself to the least amount of therapy and theory. Carry an Occam's therapy razor in your pocket. Cut out extraneous head tripping that interferes with your imagination initiating a flow experience. Reserve theory for postmortem analysis. On second thought, limit it there also and try creative analysis for a change. Do the least. Again, do as little as necessary to get things moving forward. Then get out of the way and let things move themselves forward. Quiet your mind and let the emergent mind of therapy do the talking. It is wiser and less educated than you.

No psychology: The professionalized rituals of thinking and treatment called "psychology" should be silenced. Anything that further concretizes the problem is not allowed in the room. This definitely means dispensing or undermining the presence of any Diagnostic and Statistical Manual of Mental Disorders (DSM) handbook. If you have one in your office, paste a new cover on it. I recommend a cartoon. Invite clients to bring cartoons for your manual. When they ask why, simply say that it helps remind everyone that the book is a joke. Remind them that Federico Fellini (in Barron et al.,

1997) once directed that "labels should go on suitcases, nowhere else" (p. 33). Psychologizing (and this includes family therapy talk because most of it is recycled psychology applied to a social unit)—the use of hypotheses for explaining the inside and outside workings of human beings—should also be silenced, unless you can create, reframe, or transform a notion whose absurdity actually sparks a creative opening.

With the intellectual acknowledgment of complexity, originally posed by literary contributors from Fyodor Dostoevsky to Joanne Greenberg and later joined by the complexity sciences (the first of which was cybernetics), we were able to go beyond the boundaries of our skin and see the primacy of relationship and interaction over psyche, psychology, and individuality. This is not to say relationship without individuality. Rather, there is a yin-yang here where individuality and relationship are a dynamic complementarity. Here, each dancer's steps are accounted for by reference to his or her dance without adding any unnecessary multiplication of explanatory principles.

The whole hurly-burly world of interactive relational process is missed when we are blinded by inappropriately concretized metaphors. The psychological/psychiatric blind spot does not see that it does not see interactive, relational phenomena. We become blind to seeing that an "individual" is a metaphor for a virtual transient nodal point within ongoing processes of interaction in a complex ecology of relationships.

No sociology, philosophy, theology, anthropology, or mythology; no social science, medical science, pharmacological science, or political science; no modernism, premodernism, or postmodernism. In other words, no -ology, no -ience, and no -ism: Keep them out of the room, unless you are skilled at creatively playing with them and making them transformative. If talking about cultural narratives, political oppression, epistemological errors, medical explanations, scientific studies, Greek gods and goddesses, cosmological creation stories, or theories of drama adds more quicksand to Act I, then set up a blockade that does not allow their entry. It goes without saying that if a client brings up a metaphor or theme and you can use it to creatively move things forward, then go with it!

No particular narrative: All theory and talk that frames or provides stories or meta-stories with presupposed authority and legitimacy are potentially an imposition of a semantic cage or prison of meaning. This is so even if they purport to be about freedom. This includes my own notions, which is why I am doing everything I can to circumscribe them with circularity, paradox, and absurdity. Pardon me if my epistemological slip is showing.

Clients also are not to be seen as walking around with impoverished stories to be edited by a therapeutic copy editor. I am not saying that stories shouldn't be part of therapy. They simply don't have to be framed by a narrative ideology. In other words, the use of stories as a required component of a schooled approach in therapy is negated, whereas any good yarn is worth a spin around any clinical block.

Don't be organized by therapeutic fashion: Follow the client's lead, rather than the social fashions of the therapy ideological marketplace. For instance, don't try to look postmodern, or systemic, or whatever, to avoid looking like a client-centered therapist or an intergenerational family therapist or any other well-established orientation that your colleagues may think is passé. Be what is called forth by the particularities and integrity of the situation, even if it means acting like an old-fashioned therapist or as a cowhand, bartender, or hairdresser.

Improvisation is the thing: Every therapist is required to be improvisational because clients don't always read from a script. Every moment requires being ready to "wing it." In spite of this fact, schools of therapy prescribe fantasized conversations that both clients and therapists are supposed to enact. The therapist may be rehearsed for such action, but unfortunately, no one preps the clients to say the right lines. Thus, the schooled approaches must often sit on the bank and patiently wait for the client to say the right thing before the next step can progress. For instance, a problem-solving therapist can't go anywhere unless the patient provides a clearly defined problem.

When therapists do not try to strictly follow a script, they are freer to improvise and be moved by the unique particularities of what the situation is calling for. In this case, neither therapist nor client has a clue where they will go. As improvisers, our job is not always to come up with clever lines. Paraphrasing Del Close, one of the masters of improvisational theatre, our job as improvisers is to "make our [clients'] shitty lines sound good" (cited by Nachmanovitch, 2005). Our art is similar to improvisational theatre. The difference is that therapy is a reality fantasy of life rather than the fantasy reality of the Broadway stage. Or is it the other way around?

Act in order to see, hear, and feel: Rather than first know or diagnose in order to know how to act, the creative therapist jumps in with action and sees what happens. Find out what a client does in response to your effort to transform him or her. That is an *interactional diagnosis,* based on acting in order to know what you next have to work with. It also defines knowing as creatively relational rather than as reproducing certain kinds of explanations, interpretations, and prescriptions. In this therapeutic flux, an immediate outcome shapes and directs the subsequent move by the therapist. The client's responses direct the therapist's conduct. Sitting back and doing nothing, or not intervening, may be seen as disrespectful of the client's directions to be directed.

Mirror or echo back the absurdity of the client's situation: The premises, beliefs, or actions that organize a client's situation can be mirrored or echoed back but should be exaggerated enough (given enough "noise") to set up an experiential "no exit" where the situation can neither be affirmed nor negated. In this paradoxical zone, a creative leap must be invented in order to transcend the either-or oscillations between rejecting and accepting the ongoing situation. In this Jujitsu-like way, a creative turn may be born.

In World War II, a strategy was implemented in Southeast Asia in which radio broadcasts sent messages about the enemy's beliefs, but with an exaggeration of around 30%. The art of construing an effective creative double bind is that it cannot be too exaggerated or else it will be easily dismissed as mere absurdity. It also can't be underexaggerated because dissonance and tension need to be felt. In the no-exit-except-for-creativity zone, an out-of-the-box leap is brought forth. The situation generates the tension necessary for the emergence of a creative alternative. This idea has relevance to precipitating transformative moments in creative therapy.

Get "shipwrecked": Borrowing a quote from Eugen Herrigel's classic book *Zen in the Art of Archery* (which was mistakenly not Zen, but Daishakyodo), I propose that what we unknowingly seek in clinical sessions requires being "shipwrecked by our own efforts" so that life throws us an unexpected "lifebelt." Those of us wanting creativity must be ready for an "accident" to happen—which, according to composer Igor Stravinsky, is the "only thing that can inspire us." As novelist Pamela Travers (in Barron et al., 1997) eloquently develops this idea, the creative accident is nothing less than encountering the "Unknown," which is "not so much to be understood but stood under while it rains upon us" (p. 42). Poetically speaking, our aim is to facilitate these accidents and fall into the unknown, thereby becoming midwives of creative therapeutic transformation.

"Not-knowing" is another way of referring to this availability to the unknown or the creative unconscious. As Edgar Degas (cited in Goldwater, 1945) proposed, "Only when he no longer knows what he is doing does the painter do good things" (p. 6). The same can be said for not-understanding, not-diagnosing, not-intervening, not-strategizing, not-empathizing, not-listening, not-conversing, not-storying, not-caring, and not-doing-therapy. These negations serve to free us from any cognitive stranglehold that impedes creative participation. They paradoxically help move us from "mindless doing" to more "mindful being." Our goal is to transform "doing therapy" into "being therapeutic."

The loosening of any hold by what and how we behold makes us more available to the unpredictable entry of the creative winds. When we are not attached to a therapy model or schooled approach, we are more able to allow multiple perspectives, differences, positions, hypotheses, and clinical intuitions to pass through. In this state of nonattached clinical mind, there are no fixed positions—only occasional breezes of one passing idea after another. Here the productions of mind are not sticky; they move on, making room for the next creation.

Ask what a creative filmmaker would do: Imagine a filmmaker accidentally walking into a therapy session and mistakenly thinking it was a job interview. Previously told that she should come for a creative job interview, she is ready to demonstrate a plan to shoot a documentary of the very people who interview her. She somehow gets the wrong address and enters a clinic. When she enters the session, she listens and then states, "I am a filmmaker. What could you do this week that would be interesting enough for me to film and show the world?" When the clients realize she actually is a filmmaker, they start to offer various ideas that make themselves interesting and exciting to an audience. I don't need to finish the story. It is obvious that making art with people is more transformative than therapeutic posturing.

A creative therapist knows what to do and spontaneously does it: Therapy embodies an immediacy of perception and action. Therapeutic action does not solely arise from reasoning, but from an immediate and effortless soulful release of expression triggered by the situation. The situation should be honored for bringing forth expression. Let the situation do the important work. Your responsibility requires diligent attention to blocking your temptation to construe a delusion of understanding along with any logically deduced treatment plans. In other words, you must work very hard to keep yourself from interfering with the natural process of transformation that is already taking place before you have thought an idea or expressed a word.

Clinical reality is perceiver-dependent: This is not because the therapist "constructs" it as he or she pleases, but because what counts as real is inseparable from the presence of the perceiver. Seeing pathology in a client tells us as much about the therapist as the client, probably more. Likewise, seeing beauty, hope, comedy, and absurdity is inseparable from the therapist's habits of punctuating therapeutic interaction. What you see is a consequence of how you invent your presence and interaction with the other.

A school of therapy is nothing more than a snapshot of the person who invented it: All schools of therapy from strategic family therapy to psychoanalysis and client-centered counseling are theoretical rationales for the natural conduct of the person who invented it. Anyone who tries to follow a schooled approach is like a stage impersonator who mimics other stars. If you know Salvador Minuchin, it's no surprise that he advocates restructuring a family. Similarly, if you ever met Virginia Satir, it's not surprising that she mothered her clients. If you don't know Minuchin or Satir, then read their clinical theories. That will tell you about *them*. Because you ideally want to create your own individual style of therapy, examine your own presence in life. Then utilize it. Only you can be a master of all the therapeutic gifts you bring to a session.

Don't be intimidated by statistics and outcome studies; they are highly overrated: One of the biggest commercial miscommunications in today's culture is selling the idea that a statistical outcome study has anything to do with fundamental science or "proof" that something is true. Statistics proves nothing. At best, it may help you decide to discard an improbable proposition. Please know that there are strong arguments against the relevance, validity, and assumptions associated with statistical outcome studies in social studies (Bakan, 1967).

Political posturing that asks for outcome-based treatment in today's clinical climate should be challenged. It deserves to be critiqued and asked whether it is out of step with therapy as a transformative art. Therapy is more like the activity of a stand-up comic or Broadway show than it is about predicting agricultural crop yields. Would anyone conduct an outcome study on the effects of reading a particular poem? Before answering that question, consider whether it is a ludicrous thing to do. How is an outcome study of what people say in a therapeutic hour any different? We have to create so many levels of abstraction to make the experimental design work, resulting in a situation where we have no clue whether our fantasized experimental situation has any relationship to the experience of being in therapy. By the way, I know that all this is true because I imagined conducting a double-blind, double-deaf, double-brain-dead, and double-bind experiment.

"Not-doing" doesn't look like you are not doing anything; it often looks active and fully engaged: Some therapists make a mistake in handling abstraction when they profess that "not-doing" in therapy means not looking like you are giving advice, not expressing an opinion, or not intervening. If the situation calls for these actions, they must be expressed. It's the lack of premeditated action and premeditated inaction that interferes with the spontaneity of *natural* practice. Nothing is more rigidly imposing and awkwardly purposeful than the acting out of any form of political correctness, whether it is motivated by prejudice for or against (is there a difference?) gender, culture, science, politics, religion, or preferences for room temperature.

"Not-doing" may bring forth aggression: The idea that therapists must always smile and be polite comes from a theory that commits violence upon the complexity of human experience. Aggression is usually more culturally unacceptable than sex, yet, as Whitaker (1975) proposes, "constructive aggression by the therapist is one way of defining his integrity as a person" (p. 2). Sneers, snarls, poker faces, boredom, and inappropriate laughter, among other lightweight versions of aggression, should be expressed when the situation naturally calls them forth. As Aristotle reminds us: "Anyone can become angry—that is easy, but to be angry with the right person, to the right degree, at the right time, for the right purpose, and in the right way—this is not easy."

Clients are not as fragile as professionals assume: If we assume clients are unable to handle certain interpretations and actions, then we set in motion that self-verifying interaction. If we congruently address their strengths and resiliency, they will activate expressions that match. If we are tuned into the mind of therapy, the situation will call us to be as strong, gentle, loud, or soft as is appropriate.

There is only the present, and the past is a construction of the present: All you need to know about the past is taking place in the present. When clients talk about the past, it is their way of being present. If they don't invite you to go backward in time, then don't. Accept—that is, utilize—the temporal frames they present as long as they point the way to a more creative interaction. Do not believe that anyone has a reliable time machine and therefore lean toward staying in the present.

A therapist should be willing to expose his or her own craziness and nonrationality: Whether it be speaking nonsense, making an unannounced exit from the room, producing an odd sound, taking a nap, asking a question that should have been addressed to another client, moving the trash can next to the window, eating a piece of a tissue, scratching the wall, barking out loud, or dancing on the table, the situation may call upon deep-mind expressions that will not be immediately (or ever) consciously understood by either client or therapist.

Don't share your major personal problems, parts of you that need help, or your own family struggles: Only share fragments of your problems, bits of pathology, samples of your own eccentricity, glimpses of free associations, and edited unedited fantasies. Exercise wisdom. When something needs to be exposed but requires editing, then present it in code. If you are upset about a personal issue and it is affecting a session, then say nothing more than, "I'm sorry I may seem off to you today; I'm in the midst of a personal hassle that has nothing to do with us but is distracting me right now."

If you think you know what to do with a client, you don't, so don't do that: Conscious knowing and purpose are highly overrated. Trust the whisper of intuition to lead you to the unknown. Trust other whispers that may silence, alter, or amplify the previous whisper. Allow yourself to get lost in a session and not know the where, why, or when of the moment. Allow the lost moment to get things moving.

Assume that all problems and symptoms were elected to office: The relationships that comprise a system unconsciously chose both the symptom and symptom carrier. The family chose the daughter to eat less; the couple elected one member to have an affair for both of them; and the grandchildren nominated a dead grandparent to stir things up. Respond to these public elections with respect and proceed to make a contribution

to their complexity. Perhaps you will be called to invite all family generations and neighbors and lovers to a session, order lunch for an anorectic,
or place an apple on top of a catatonic's head and promise him you won't
tell William Tell.

Use action koans to help break stuck patterns: Encourage clients to
enact a parody of their own situations, doing so within the frame of an
exaggerated fantasy. For instance, have an insomniac wake up three times
a night to listen to a recording of his or her own snoring, or have a worrying mother hang her adolescent's underwear in a tree whenever the child
comes home late, or write a prescription for taking a raisin before and after
each problematic argument. The rational leaning Tower of Pisa won't fall
unless it gets tall enough. Add more babel, particularly of the absurd kind,
to any overdiscussed life issue but avoid more rational discussion.

If it at first seems like chaos, then something may be changing: If it's
familiar, nothing will change. The motor for transformation has been called
anxiety, confusion, and chaos. Honor those experiences as evidence that
the process the client both desires and fears has come alive. Second-order
change is unfamiliar to the eyes of first-order change. "Out-of-the-box"
means it is outside of one's habituated recognizing and understanding.

In other words, "this is a way of saying that the therapeutic problem
is to increase the complexity of the situation rather than restore order"
(Whitaker, 1975, p. 5). Like the development of mystics, beginning therapists start at a state where they hate ambiguity, then move on to tolerating
it, later welcoming it, and finally seeking it out (D. Arispe, personal communication, October 15, 2008). Fully seasoned therapists may even try to
"induce chaos and craziness" (Whitaker, 1975) rather than bring things
back to any status quo. They do so to help facilitate creative transformation, liberating what is unable to move forward because of how it is caged
and stuck in a restrictive, stagnant pattern. Our natural capacity for creative growth is able to express constant transformations of order, moving
from one pattern to another. For this to be set in motion (not interfered
with), we must not be overattached to any particular obstacle.

Disorder, another name for the unknown or unrecognized, emanci-
pates, spawns, and calls forth new order. One of the founders of cybernetics, Heinz von Foerster (2002), expressed this paradoxical principal as
"order-from-noise." Through noisy disordered chaos we nourish the processes of transformation and thereby help generate the new. We give birth
to what we are looking for through a journey of being lost.

This does not mean that we need to have a new marital partner each week,
frequently trade in our children for upgraded models, annually change
careers, or periodically move our residence. It more likely means the opposite. To have more stability in our lives, we require constant change. In the
voice of old-fashioned therapy talk, individuals need to change and grow in

order to conserve the integrity of their relationships. To avoid falling over while standing in a canoe, you must rock back and forth. Movement helps stabilize your steady position. Relationship systems, whether in the home or profession, need to become a finely conducted symphony that includes interspersed disagreements, arguments, and discord in order to maintain a general sense of overall stability, noise, uncertainty, and deep meaning.

It takes great discipline to be wild and chaotic in a transformative way: A creative therapist must not engage in the outpouring of raw unedited primary process gibberish. Nor should she be overly rational and neurotically precise. The art of having spontaneous wise expression pulled from you requires great preparation. Some of the preparation is hard study and disciplined devotion. There is no easy path to becoming therapeutic. It takes a lot of advance work to be effortless in a single session. The samurai must give his life to becoming a master of technique and swordsmanship. Then and only then can the technique and know-how be thrown away so that the sword and swordsman are neither one nor two.

Consider pretending that you know when you don't and pretend that you don't know when you do know: This double reversal may help spring forth a creative tension in the room. Experiment with this posturing and then try it in other domains of experience. For example, express great passion for that which you didn't know you were passionate about and less enthusiasm about those things you are worked up about. It is likely that somewhere within your deep unconscious you are interested in that which on the surface seems to be a disinterest. That disinterest could not be present on top without its opposite buried below. Enact the opposite to wake up its presence in a session. Use this bipolar method of deep-sea fishing to find new ways to jump-start your creative processes. Be careful about your preferences. "The truth which is important is not a truth of preference, it's a truth of complexity … the dance of Shiva" (Bateson in Brand, 1974, p. 32).

Allow yourself to ask a client to do anything different before you begin: Even be ready to ask your clients to take a 15-minute walk around the neighborhood before you start the session. Tell them to *imagine* that the change they desire took place 1 hour ago and that they will start to realize it in the next 15 minutes. As they imagine this happening, invite them to bring back an object they find on their walk, perhaps a stone, leaf, twig, bottle, or piece of paper, that is symbolic of the change that has started. Now creatively work with what they present.

Only one person can occupy the same craziness at the same time: When someone is really acting crazy, be ready to become crazier than they are. John Rosen, the innovator of Direct Analysis, once visited a clinic where I worked and was told that a patient believed he was the Pope. Dr. Rosen immediately called a taxi and had the patient pack his suitcase. He took the patient to the airport in a rush, saying that he had just heard on the radio that the Pope was giving a talk at the Vatican the very next day. The patient, responded, "Doc, you're crazy." Sometimes we will be called to be crazier than our clients as a means of helping push them out of their own stuck way of being in an impossible position.

Remember the jet-propelled couch? A famous physicist at Los Alamos lived in a delusional world where he believed extraterrestrial beings were sending him secret scientific information. The psychoanalyst called in for treatment decided to learn about the scientist's crazy world and eventually started to believe it was true, staying up late at night trying to decipher various star maps. The more lost he got in the physicist's fantasy universe, the more the physicist began to bring the psychiatrist back to earth. It's worth noting that social interaction, rather than psychotropic drugs, brought both of them back with a safe landing.

What we have to work with in therapy is metaphor: We need to be careful not to underestimate the unique way metaphor voices the complex truths of therapy. Metaphors cannot be clearly defined; they must hang suspended between opposite meanings. The metaphors of therapy cannot be quantified or measured. They are more like a shape than a size. Bateson and I once created an historic conference for mental health that was hosted by the Menninger Foundation in Topeka, Kansas. We called the conference, "Size and Shape in Mental Health" to make this very same point (Keeney, 1983b).

Most of the therapy profession historically went the way of thinking that what it did belonged more to the world of "size"—inhabited by static nouns that signify thing-like substances that could be subjected to quantification and measurement. Here, no dancing verbs or slippery metaphors are admitted to therapy, only reified variables that typically belong to "sickness" or "treatment." This is where we get lost in a linguistic wonderland; we hunt for Alice-in-Wonderland abstractions that we presume are causative of pathology and search for molecules or treatment plans that lead to cure. This "hunting for snarks" is as schizophrenic as a person eating a menu rather than food or a therapist implicitly asking a client to enact the therapist's theory. All in all, therapy too frequently makes mistakes in how it handles abstraction.

The other choice for therapy is to follow the road signs that point to metaphors of "shape." In doing this, we find pattern, communication, interaction, circularity, and the relational and more feminine dance of therapeutic flux. This domain does not fight against the paradoxical, slippery, ever-changing nature of how we know the world *and* how we are in the world. It accepts nonmeasurable slipperiness as our epistemological and ontological position and proceeds creatively to engage, relating in transformative ways that are inspired by a wisdom that arises from the dance, rather than from a single dancer assumed to be operating alone.

If a client says he is from Mars, we do not have to think that this is evidence for madness. Instead we can assume that it is a creative rendering of a complex truth carried by that choice of metaphor. Perhaps he is like a cheap candy bar that everyone feels free to take a bite out of, or he feels like others have just discovered water (or tears) on his surface, or that he is only up when everyone else is down. Creative therapists accept metaphor as the language that speaks of and for transformation.

There are only vicious circles and virtuous circles: What psychiatry regards as "symptoms" are simply different forms of vicious circles. Sadness about sadness is a vicious circle, inducing an immobilizing heaviness called "depression." Feeling out of control about feeling out of control is a spinning called "anxiety." The way out of one circle is to start another circle, one that is virtuous. Joy about joy may virtuously cycle into ecstasy, giggling about giggling inspires uplifting laughter, and movement that accelerates movement breeds growth. Finally, initiating a creative action that, in turn, inspires continuing creative responses is the circular-recursive road to therapeutic transformation.

If you still think you have to recommend a pharmacological prescription, then medicate everyone, including the psychiatrist: The interactive relational framing of psychiatry can be defined as a situation in which an individual is brought to a doctor because someone is complaining about the individual's nonconforming conduct, perhaps an employer, a spouse, parents, friends, or even the client. Someone is uncomfortable with the unpredictable, uncontrollable conduct of a less conforming member of society. The so-called "problem person," or patient, also makes the psychiatrist anxious. This disturbance, in fact, stirs great anxiety in the whole social system where the patient's drama is performed. The psychiatrist proceeds to convince herself that this is indeed a serious problem and then proceeds to medicate the patient, perhaps giving him an anxiety pill. The psychiatrist's anxiety, along with the anxieties of all the people complaining, immediately subsides. The same outcome would have taken place had the psychiatrist, rather than the patient, taken the medication. Better yet, the same outcome would be achieved by medicating everyone but the patient.

In the early days of systemically oriented family therapy, families' demands that their identified problem carrier be medicated might be countered by a therapist saying, "I'll medicate Bob, but only if you *all* take the medicine." Similarly, in places like the Philadelphia Child Guidance Clinic, it was not unheard of for whole families to be put in the hospital rather than one individual. Family therapy, in its beginning, already knew what contemporary ecologists are learning today. They recognized that there were no individuals independent of context—that is, no one exists outside of their interactions with another. There simply is no such thing as an individual without a contextual home.

The basic unit of therapy is **relationship,** ***not individuality:*** This may be the most important idea a therapist can embrace. When it is realized and incorporated in therapeutic practice, therapy becomes a dance rather than a doing-something-to-another. When we are primarily concerned with the dancing rather than either individual dancer, therapy becomes a moving art. By "moving" I mean that it is always changing and that change inspires more change. Creativity, too, is about relationship and is concerned with *the relating* rather than the people or things that are being related. When we deceive ourselves into thinking that we are separate, the illusion of unilateral control is fostered and creative interaction is stilled.

The interaction between therapist and client is the locale of transformation in therapy. It does not take place within the client and it is not caused by the unilateral action of a therapist. Transformation takes place in the dynamic interface of client and therapist, what I once called the "mind of therapy" (Keeney & Ross, 1985). When therapist and client come together, symbolically walk together, and then metaphorically dance together, a mind greater than either component is co-created. The voice of transformation is not uttered by the psychological mind of a therapist (or client), but by the wider mind of therapeutic interaction that embraces, connects, and transforms all who participate.

William Irwin Thompson (1977) proposes that we live "at an interface between opposites: earth and sky, sea and shore, life and death. … Yet it is precisely the interface between opposites that is the place of transformation, and the energy of that transformation comes from remaining poised at the perilous edge" (p. 49). Transposed to therapy, we may say that transformation comes from our being poised on the perilous edge of our interactions, the dancing of our relating with one another.

When the concept of self diminishes its importance as a nodal point in our experience, we find it easier to recognize relationship and the ecology of interactions that sponsors our experience. This higher order knowing reveals (Bateson, 1972) "a world in which personal identity merges into all the processes of relationship in some vast ecology or aesthetics of cosmic interaction" (p. 306). Here we are able to flow with creative process and be a transformative midwife to systems in need of change. In relational knowing, we abandon sitting still with the constraints of individual mind and move inside relational mind. This more encompassing mind is not harnessed by the narrow sight of conscious mind, the limited partial arc of the whole circle of relational mind. Relational mind carries wisdom not accessible to partial mind.

Never forget that there is no such thing as therapy: Develop a sense of benign absurdity about the whole profession, knowing that it is the art of being impossible in order to dance with the impossible. Doing less sometimes leads to *being* more. Therapy can be as crazy as it seems and, if you think about it, you will go crazy.

Don't underestimate how seductive theory, narration, and interpretation can be: As beautiful and complex as highly abstract theories like Jungian analysis and postmodern diatribes are, to cite a few examples, they arguably have contributed less to evoking transformation in a therapeutic encounter. Beware the temptation to overinterpret, spinning an elaborate web of new understanding while going nowhere. Of course, moving creative sessions have been conducted by Jungian (and modern and postmodern) therapists in spite of their theoretical weight. It most likely took place when they stopped being interpretive and indulged in a moment of play not blinded by their theoretical shadow or they were tripped by an accidental deconstruction of deconstructionism.

Embody opposites: The creative therapist is free to embody contradictions, ambiguities, and opposites. Without the tension between contraries, there would be no creativity. Consider this insight to be a way of honoring the sacred. Jung (1973) says this nicely in his *Answer to Job*: "All opposites are of God; therefore man must bend to this burden; and in so doing he finds that God in his 'oppositeness' has taken possession of him, incarnated himself in him. He becomes a vessel filled with divine conflict" (p. 65). The creative therapist is nothing less or more than a vessel holding myriad contradictions and conflicts. Utilize them to activate transformation. As Walt Whitman (1900) asked and answered himself: "Do I contradict myself? Very well then I contradict myself. I am large, I contain multitudes" (p. 3).

Between the contraries we find our existential truths. They enable us to express the kind of wisdom familiar to poets like William Blake. Take this example, spoken in an actual clinical session: "You must find someone you disagree with, a worthy enemy, so you can be differentiated from what you are not." For existential therapists, this advanced social scenario is called "marriage." At the same time, in the interactive dances that differentiate us, you and I may truly become a "we." In turn, the circular rounds of relationship bring us each forth. As von Foerster (in Franchi, Güzeldere, & Minch, 1995) summarizes this emergence: "In becoming 'we,' each of us realizes 'I'" (p. 1).

Yes, what I am advocating—complexity, not-doing, and creative therapeutic transformation—is at odds with our instrumental, control-obsessed culture: Any work that aims to transform our therapy, ourselves, and our clients is also a way of changing the cultural premises that organize the other institutions and performances of everyday life. Becoming more creative in any domain may bring about disorder in all forms of the status quo, whether it addresses clinics, public schools, neighborhood church meetings, or environmental policy. What we do in our sessions is inseparable from how we relate and interact with the bigger problems of our time. Remember: it's all connected! The flap of one therapy bread-and-butter-fly's wing could cause a storm in the capitol.

Should we dare address whether the helping professions contribute to endangering the sustainability of contexts holding mental and biological health? I believe we cannot afford to avoid asking the same kind of question that was voiced by Rachel Carson's concern over the way agribusiness handled bugs. Shouldn't we, as clinicians, be as concerned over how we handle the pests and symptoms of everyday living? Are the medically prescribed chemicals going into bloodstreams as risky as the unexpected side effects found with the use of insecticides? Is the *thinking* behind the use of certain chemical, political, and clinical interventions a greater pathology? Should we at least ask the question?

Several years ago, the lead article in the *Journal of Marital and Family Therapy* (Johnson, 2001) posed an answer, arguing that Bateson, von Bertalanffy, and other leading thinkers of systems and ecological thinking had been fanatical and erroneous in their outcries about the relevance of our ecological crisis to therapy. Much like the early critics of Rachel Carson, the author minimized and ridiculed the importance of there being any concern about how the premises we embody in clinical practice may be inseparable from how we affect our ecosystem. (Perhaps profits are more valued than prophets of the silenced springs?)

This same contributor became the president of the American Association of Marriage and Family Therapy, the mental health organization that published the journal. What happened to the field of family therapy, which began as a prophetic outcry for systemic understanding, relational sensitivity, ecological responsibility, and a respect for comparing the thinking that is relevant for the planet with the embodied ideas driving a clinical conversation? Has family therapy lost its systemic soul?

More generally, the creative and revolutionary spirit of the original founders of the whole of psychotherapy, including Fritz Perls, Viktor Frankl, and Virginia Satir, seem less present in today's clinical world than the resurgence of standardized texts in graduate education, multiple-choice licensing exams, an emphasis on prescribing a biomedical view, a backslide into myths of understanding that are timid to act, and quantifiable outcome studies, all marching side by side with the profiteers of agribusiness, pharmacology, and professional unions. Has therapy also lost its creative soul?

Because I was one of the three systems theorists, along with Gregory Bateson and von Bertalanffy, critiqued in the aforementioned essay, I say now, even more passionately than I did the first time I said it more than 25 years ago: therapists should urgently reconsider the ecological and systemic critiques of our wider cultural ways of thinking and seriously ponder whether we are part of the problem. Not only with respect to global warming, overpopulation, and the destruction of our natural resources, but to the way the same kind of thinking threatens the sustainability of our being a healing presence within the mental health professions.

With respect to whether this plays itself out in clinical work, let us ask whether therapy has become significantly less contextual, less systemic, less holistic, less relational, less feminine, less interactional, less complex, and less inclusive of diverse cultural modes of expression. Are the mental health organizations and institutions pushing us to be more organized by simple reductionist models and lazy stereotypes, thereby making us more trivial and predictable? Are the complicated theories of therapy little more than a complication of simple models? Have we lost a respect for the complexity of therapy? Are we losing sight of the "art" of therapy? Have we all but killed the presence of creativity in therapy? Do we need to address greater change and a larger transformation than most have imagined? In particular, do we need a therapy of psychotherapy? Perhaps our very way of knowing, our clinical epistemology, must finally get serious and join the creative (and necessarily ecological) revolution!

On the other hand, it's less likely that you'll be eaten by cannibals if you are not a missionary: We need to be clever as a fox and wise as an owl in our interactions with our clients, profession, and culture at large. Get yourself inside a Trojan horse, enter the enemy's fortress, and then offer an unexpected surprise. Be a double agent inside the belly of each session and inside the guts of this profession. Become a nontherapist disguised as a therapist. Doing so will feed creative transformation with the inevitable tension it holds.

Do the least harm and foster the most charm: I am utterly flabbergasted when someone asks whether absurdity, or an assigned task, or a strong statement in a session might "hurt" a client. The same person usually has no issue with the use of psychiatric diagnostic stereotyping and heavy doses of pharmacological intervention into a bloodstream. Which action would cause the least harm: burning a lampshade or taking an antidepressant? (I once advised a client to do the former, but never the latter.) I believe that the interventions of psychiatry are the most invasive and should not be considered until all other ways of trying to make a difference have been tried. Let us first begin with creative interaction before mindlessly dealing pills, electrically stimulating brains, or locking patients up. Clients are better served when we bring more charm, enchantment, and mystery to their lives—all of which whirl the winds of creative transformation.

Don't listen to what I have to say; listen to what each session is asking you to become: Whatever I say is only intended to loosen, provoke, evoke, and help free you from whatever holds you back from expressing creative transformation. Words bind as well as set us free. Be less rational with what you relate to therapy and more therapeutic with what you think about relating. The way out of therapeutic impasses is through entering the unknown and unpredictable processes of transformation, which may be best accessed by abandoning what consensually looks like therapy. As Miles Davis (in Barron et al., 1997) said, "If I ever feel I am getting to the point where I'm playing it safe, I'll stop. That's all I can tell you about how I plan for the future" (p. 56). Be available to be called forth by the unknown, the unconscious mind of therapy, and the invisible heart of transformation. It is here that we find "the highest aspiration … to be responsive to the needs of the particular situation" (Varela, 1999, p. 71).

The British novelist Pat Barker (2008) was asked if the psychological damage inflicted on families due to war (in this case, World War I) could be healed. Her response applies to all wounded human beings: "I think it can. And I don't think it necessarily requires professionals. It requires someone to do something very, very creative with the materials of everyday life. A lot of ordinary people are capable of that" (p. 41).

PART **II**
Awakening a Session

CHAPTER 4

Man of Plenty

This session was conducted at a university teaching clinic in the southern part of the United States. As is usual when I teach, the student trainee (T) begins the session while a group of therapists observe from behind a one-way mirror. After a few minutes into the session, I come in to demonstrate how to awaken a session and bring forth a whole therapy in a minimum of three or more acts. In this case, the client (C) has just come from court, where his divorce was made final. He also just finished moving out of his apartment.

All the sessions presented in this section of the book are transcribed to bring you as close as possible to an unfiltered peek at the action that took place. I will provide commentary that indicates the movement that is taking place in the session. The case—the interactive dance between client and therapist—is the teaching. Please note that the following transcribed session is track 1 on the enclosed DVD.

Act I: Divorce

T: So you moved?

C: Yeah … today was the day that I moved. Last night was the night that I moved all of the food over to the refrigerator … and so it's come full circle because this morning at 9:30 I was in divorce court and my divorce is now final.

T: I was going to ease into that one.

C: (Laughing) No easing into it today, it's been there all day for me. A friend was my witness. I went and my ex did not show up … and that was the fear. I was afraid that she was going to show up. I didn't want her to show up. Because I knew that it would be too emotional for me because …

T: Yeah? …

C: … the gentleman that she's engaged to—they are going to … get a house together and they are already living together. They are going to get married and he's going to quit his job at the hospital and do what ever she's doing. That's what he has told other people at the hospital that have told me.

T: Well, she's moving on, you know.

C: And I'm not …

T: What do you want to do?

C: I want to move forward.

T: How?

C: (Laughing) See I say that in one breath and in the other breath I'm fixing to say is … (C sighs.) … I'm going to go out to dinner tonight with a friend from work.

T: Now is this with …

C: You're a bit confused.

T: I'm a bit confused with all of the ladies you've got going on …

C: This isn't anybody that I've ever talked about.

T: Okay, so this is a new lady …

C: I've not even talked about her with you.

T: Okay, all right, just a new lady.

C: She's just a girl that I work with …

Act II: Beginning or End?

(Brad Keeney [BK] knocks on door.)

T: Come on in.

BK: Hi.

C: Hi.

BK: I'm Brad.

C: Brad, nice to meet you. I'm Rob.

BK: Nice to meet you. I just had a chance to listen to a little bit of the conversation. And I'm not sure whether we're at the beginning or at the end. Is this the end of your therapy?

I introduced the distinction of "beginning or end" to see what he connects with it—the divorce, therapy, his new friend, or something else.

T: This is the end. We actually played poker last week and we were betting the number of sessions he would still come, and I whittled him down to two. So you are one before the last.

C: But I'm looking more for the deck of cards right now cause I'm thinking I need more.

We see he is having difficulty leaving both his former wife and his present therapist—something that may or may not be useful as the session
progresses. We can now address leaving therapy as a metaphorical way of
addressing how he leaves others. Whether this develops into a theme is yet
to be determined. We will see what moves forward.

BK: So why would you have one more session after today?

T: Well, the last session we had always planned would be a celebration …
about how this is the transition, a change from his life. … So
today just happens to be kinda smoothin' the way for that.

BK: So you're done? … I'm not saying your life's going to not have more
challenges, and there's going to be ups and downs, and all the
stuff that's part of everyone's life …

C: No, it's so comfortable. It's so comfortable.

BK: You've grown accustomed to it.

C: Yes.

BK: You've got a nice friend here.

C: It's so easy, it's so, I … As I sit here and talk about my ex-wife like I
just did, to him, he doesn't look at me and say to me, "You're an
idiot."

T: I have been hard on you, though.

C: Yeah, you have been hard on me, but not as hard and as cruel as [my
other friends].

BK: (To T) So I'm not so certain … you've … been such a good therapist.
He can't let go of you. (C and T laugh.)

T: See that's the thing, I needed to be harder with him.

BK: I don't know. He likes when you're tough, too.

T: Yeah.

BK: You like your therapist?

C: Uh, yeah.

BK: You really like him, yeah?

C: I enjoy him.

BK: You enjoy looking forward to him?

C: Yes. It's, it's, it's a time to talk about all the emotions that's been
going on, it's … this is really gonna sound crazy, but it's like a
relationship.

BK: Absolutely. So, maybe it's a day for two divorces.

C: Oh, God. Gee thanks.

BK: Double whammy. You get to have twice the party tonight.

C: (Sighs) One of the things I was going to talk about was …

BK: (Interrupting, moving his chair closer) No, no … I mean this because, last thing I wanna see happen is you become a therapy addict. It's someone who likes therapy so much that they can't let it go. Cause it's a cozy kind of relationship. It's nice to come and talk to someone who's sorta like a friend but really isn't so you get to have the benefits of the sorta like a friendship but none of the disadvantages of really being a friendship, and you can always leave if you don't like it, and you get to come to it when you do, you know …

Therapy is defined as a pretend friendship that is easy to get attached to without the challenges of an actual intimate relationship. This opens the door to discussing the nature of his relationships. Note that my interruption was implicitly defined as helping stop him from becoming a therapy addict. It is also done to highlight and utilize what has already been brought up.

T: I wish …

BK: It's one of those kind of things. … This guy is actually going to graduate, too, so you'd have to find someone else. … Have you been in therapy before, [or] is this your first time?

C: When … my parents got divorced when I was 15, I went, for about … a year … I blamed myself for my parents divorce. I turned out to be an okay kid, I guess. I didn't end up being an axe murderer or anything like that … but I don't remember liking it. I remember thinking it was something that I had to do.

BK: But you like this. You like what you've had here.

C: Yeah.

BK: Yeah? So you want to be a therapist now?

This question presents what screenwriters call a "turning point" or "plot point," a different direction for the conversation to follow. I am inquiring whether his experience with therapy has taught him to be therapeutic with others. This, again, points us toward his relationships.

C: Sure, why not?

BK: Well, I was wondering. (To T) You ever have the fantasy that he is actually liking therapy so much that he would want to become a therapist?

T: No, he loves his job too much. But I think I have influenced him about how he does his job. So I think he's a therapist in disguise, undercover.

BK: Hmm ….

C: And I believe that.

Sure enough, the client regards himself as a kind of therapist. We have turned the conversation away from his receiving therapy to exploring how he is therapeutic (one way of relating) with others.

BK: Uh-huh.

C: I think that I interact with my patients differently. I, I listen to them longer. I pick up on things, more of the holistic medicine type thing with attitudes and emotions, as well as the physical, and try to find out what is going on emotionally with them.

BK: So where do you work?

C: I'm a nurse at the hospital.

BK: Who's a secret therapist?

C: Who's the secret therapist?

BK: You're a secret therapist. The nurse is a secret therapist.

C: Maybe that's why everybody calls me!

BK: Yeah? Everybody calls you? They lay out their problems and …

C: Yeah.

T: I told you he's a therapist in disguise.

The above conversational run was encouraged as a means of further substantiating and legitimizing the client's identity as a "therapist." With this connotation firmly in place, we can ask about his expertise in creating endings.

BK: So do you know how to terminate sessions with them? How to stop?

C: They get discharged.

BK: So you don't have that responsibility. Yeah?

C: With my patients, no; with my friends, they're always there.

BK: Anyway, I was curious because when I watched behind the mirror, it looked like both a beginning and an end. And I didn't know what it was an end of or I didn't know what it was a beginning of. And certainly it's …

C: It actually is a beginning and an end.

BK: Yes.

C: When I first sat down with you, my marriage had fallen apart. I was devastated, and I fought to keep it going. I've been through a lot with it and now that my marriage is over with, that part has ended. Now the beginning part is the rest of my life.

We have defined his present situation as both a beginning and end, enabling us to question, underscore, or frame any topic as either a beginning or end, or both.

BK: Right.

C: And I need to move forward.

BK: And certain things have changed in your life. Once upon a time when you were in therapy you didn't like it. Now that you're in therapy you like it …

C: (Laughing) There's a lot of things you didn't like when you were young you know? Peas, carrots …

BK: … but you are [also] practicing therapy, you're being therapeutic, you're being a healing presence to others. So what can we do for you? … I'm only going to see you this one time. This sounds like this is the last session right before the party. What could we do today that could make a difference? I mean is there anything? Or are you just wanting to wrap things up and move on? Cause you know life's not going to get easier.

Some clinical observers may think I am acting as a problem-solving therapist here by asking what we could do for him in this session. Though several therapeutic orientations ask this question at the beginning, I am simply interested in seeing what metaphor he presents after we have established that he is inside both a beginning and an end. His divorce is both the end of one relationship and the beginning of other relationships in the future, including the date he has scheduled. The end of his therapy may be the end of his being a client and the beginning of his being more of a secret therapist.

C: No, life's going to get more difficult, I think.

BK: Maybe. That's part of it getting better.

C: I don't ever hear other people who get divorced having this feeling that I feel right now where, if she walked through that door, if she called on that cell phone and said I need you, I'd be there. After all she's done to me. Okay? She's cheated on me four times. She cheated on me with a guy at the hospital … yet my anger is not towards her. My anger is towards him. I guess I want validation that's an okay feeling, but I just don't get it. I don't get why that's an okay feeling. I don't understand why that's an okay feeling.

BK: It's not about it's okay or not okay; it's about it's crazy. It's just a crazy experience to both join together and then to separate. … It's even twice as crazy. It's not going to be something that you can understand. It's not going to be something that will be clear to you as to how it can be resolved. It's always going to be the presence of an absence, and the absence of a presence. There will always be anger that should be here but it's there. There should not be an expectation that this is going to be anything other than crazy. As all important things in life are. You know that being a nurse, yes? And you're also being a therapist.

Though he asks whether it is okay for him to still feel anger at the guy who took his wife, while not being angry with her, I define the situation as simply "crazy"—that is, not something we can understand. The last thing I want to do is get sucked into a hermeneutic black hole or sink into the quicksand of exploring understanding of any kind that keeps him in a situation where he is unable to either end or begin. Saying that "all important things in life" are not understandable actually honors the complexity of the situation.

C: Yeah … I would trade my nursing career for a stable relationship. I was married for 13 years the first time, divorced after school. Married for 4 years, divorced after RN school—that's this one, the second marriage. … I would trade my nursing license in a heartbeat and go flip burgers at McDonald's for a relationship … (pause) … for her back.

Act III: Get a Dog

BK: There are certain things we don't have control over. And certain things therapy can help us with and things therapy has no power to be anything other than an extended hand and, and uh, friendship in this moment. But I'll ask again, you know, what could we do in just the next couple of moments? Is there anything? … just a little thing that might make a difference? Is there such a thing? (C laughs.) Do you have a pet?

I move toward exposing the absurdity of his request for a stable relationship by asking whether he has a pet.

C: No.
BK: Why not?
C: I'm not a pet person.
BK: They're supposed to be the great teachers about relationships.
C: Don't want to deal with a smelly stinky thing in the house.
BK: Oh … I don't know that you're all that willing … (T laughs.) … to go through the trials and tribulations necessary to teach what you need to know about having a steady relationship.
C: I'm …
BK: I'm teasing you.
C: Okay.
BK: But no I'm not.

This is an interesting therapeutic move that is more obvious when viewed on the DVD. I tell him that I am teasing but immediately counter by saying that I am not teasing. This marks how we will work: I will tease within one frame and then circumscribe that frame as nonteasing serious work. I will mirror both the absurdity and seriousness of his issues.

C: I'm … pets okay? Dogs, dogs. You have to feed them, okay?
BK: Yeah.
C: My sister has dogs, okay, they can't afford crap but their dogs got six, eight hundred dollar pet bills.
BK: I'm sorry. I distracted you … What do you want?
C: The ability to walk up to [this woman friend and have] the ability to say to her, "Where are we going? What is our relationship?" Though I can't do that, I can typically walk up to anybody, even a complete stranger, and have a conversation.
BK: So you'd be a good therapist.

Whenever there is an opportunity to underscore his being a good therapist, I act on it. In general, at this stage, I am noting what metaphors are resourceful about his life and then underlining them when he presents information that fits.

C: She and I have a great relationship friendship.
BK: Who's this?
T: She's is a friend that has stuck with him from the very beginning … and even was his witness today in divorce court.
BK: Oh, so you've managed to have a long-term relationship that's stable. Right?
C: I wouldn't call it stable because she's …
BK: A long-term relationship?

At this point, I am underscoring his skill at maintaining long-term relationships but doing so as someone who is also intermittingly teasing him.

C: Yes, oh yeah.
BK: Okay, congratulations. (Leans over and shakes C's hand.)
C: Well, there's another woman friend, too.
BK: Congratulations. (Leans over and shakes C's hand; T laughs.)
C: But that, that's …
BK: How many others? You have a lot of relationships, I think, don't you?
C: Yes, I do, but … the intimacy part, and I don't mean sexual. The one who just sits there and will watch television with you. Or if you decide that you want to go to Branson for the weekend, it's okay to go.
BK: That's a dog.

Note that we have already seen how "cozy" he is with therapists who are pretend friends and friends who are pretend long-term relationships. With a dog, he could have a pretend long-term best friend.

C: But a dog can't go to the show with you.

BK: That's a dog. Take him to the drive-in.

C: They closed those down in Oklahoma.

BK: You're kidding. You can get a projector and put it in your backyard.

C: The ability to … know what to say, the right words to say my woman friend …

BK: But she can be your witness in your divorce, but you can't ask a simple question. Well then why don't you get a flash card—that is, hold up an index card, write down the things you're unable to say to her, and have them in your pocket when you …

The client tries to turn the conversation toward a deficiency—his inability to say the right words to women friends. Because of the interactive play we have established, it is easy to deconstruct this as a tough problem and regard it as a simple lack of cue cards. I will continue letting the air out of his inflated concerns in this teasing and absurd comedy about his double divorce that is both a beginning and an end. I will stand under (rather than understand) a man who has become a therapist (rather than a client) with many long-term relationships and tease him about whether his only real pressing issue could be solved by getting a dog. We shall see where this comic interplay, expressed within serious therapeutic work, takes us.

C: That's how we studied.

BK: Could you do that?

C: You know what, that's possible, that's doable. That's really doable.

T: Have you seen the movie "*Love Actually*?

C: No.

T: There's a guy who does that. He can't talk to the girl. He gets a bunch of cards and holds them up to her and goes through the cards.

BK: Right. And have them different colors depending on the degree of difficulty. You know the hottest, the most difficult, they come red hot … if it's strange or odd thing to say, make it purple. If it's life affirming, make it green. But, would you do that? That was simple. You're so easy …

T: I know, you came here expecting something really …

BK: You're so easy, so easy you're, you're …

C: I'm cheap.

BK: Look … you just got divorced, and you're able to smile and laugh and be happy in spite of the badness around it. That's unbelievable. That's a strong person. You have so many relationships … several … are long term …

T: He's going on a date tonight.

BK: Though you can't say goodbye here, you're going on a date tonight … and the kind of relationship you want that you don't have is so easy, you can just get a dog ... that's the kind of intimacy that's stable.

C: Everybody has told me since May 9th or 10th of last year: get a dog.

BK: Of course.

C: Get a dog.

BK: Yeah.

C: And I keep telling them, "No, I don't want a dog. I don't want a dog."

BK: You're so easy and this other thing is just a matter of getting index cards. So what would be the first index card, what would you write?

C: Where are we going?

BK: Where are we going? Okay, so, can you write that down?

C: Yeah.

BK: Are you gonna do it? Is this for the woman you're gonna see tonight?

C: No.

BK: No, somebody else. Whooh! (C laughs.) Why don't you have another woman friend with you when you talk to her? You've got such a knack for it.

C: I thought of that, but this one woman I'm worrying about is more quiet, and I don't think she would ...

BK: Is she quirky?

C: Quirky?

BK: Would she respond to something like an index card?

C: Yes.

BK: Then you should do it. For your own learning. It will help you become a better therapist.

C: You keep saying this "gonna be a therapist." He (T) did that to me twice. He switched seats; we switched seats one day.

T: Because you're a good therapist.

C: So who's charging who 10 dollars this week? I'm sorry. (Laughter)

T: Well trust me he's a bargain. 'Cause his going rate is a lot higher than this, isn't it?

BK: Absolutely. But the other thing you could do is make tape recordings. You could have tape recordings of the things you want to say, then you would be saying it, that's a step closer. You could tape record your words ... just press the play button. (C laughs.)

C: That's how my first wife and I communicated back and forth long distance. We taped recorded messages and mailed them across state.

BK: So you managed to have a relationship without being totally there, but by still having the voice of the marriage. You are a complicated and interesting guy. (C laughs.) You are fascinating.

Again, I am using every opportunity to validate his resourcefulness.

C: It's a lot deeper than this, and it's not my mother. It's my father, okay?
BK: Okay.
C: Okay, if I'm so deep and complicated, I pretend to understand, or if I'm such
        a good therapist, why did I not understand the Geico commercial?
T: Yeah?
C: Where he says pick up the phone, it's my mother … I didn't get that …
        until somebody said to me, it's because in therapy you always
        blame your mother … it's all your mother's fault.
BK: Yeah, you're liberated past the clichés.

Rather than jump into an archaeological dig to explore the history of his relationship with his mother, I benignly frame his comments about denying that his mother is related to his present difficulties. I do keep this mention of his mother in the back of my mind, not yet knowing how it may be useful.

C: It's like, okay …
BK: It takes more than that to catch your attention.
C: It's like that didn't.
BK: That wasn't very real huh? So good … we've taken care of that. You
        want to be able to say things you can't say, you can write down
        index cards, make recordings, which you already have previous
        history of expertise in.

I am bringing us back to what he is able to resourcefully accomplish, though it is in the double frame of absurdity and serious therapeutic work.

C: And the index cards …
BK: We have time. Lets have another one [i.e., something else to work on]
        … Number two.

Act IV: Man of Plenty

C: The question I asked you last week was how does a little fat white boy
        who's not handsome have 30 women call him and want to go
        out, and want to get to know him, and want to be his friend.
        How does this happen?
BK: Sounds like the title of a best-selling book.

Whenever possible, I avoid the pathological aspects of a communication and address the resourceful part. In this case, rather than discuss any view of deficiency about himself, I propose that his whole utterance could sell books. I will continue to tease him, doing so in a way that underscores his abilities rather than deficiencies.

C: (Laughs.) It should be called …

BK: If you get the whole thing down as a title it would probably be a best seller. 'Cause the only person who can answer that is you.

C: But I don't, I don't …

BK: You just don't want to share it. You don't want to share that knowledge.

C: But what is it?

BK: It's your magic. If you knew, it'd probably mess it up.

C: Well, I wish I could turn it back on then.

BK: What, you want 40?

C: I wish I could turn it back on my wife. My ex-wife.

BK: Well, maybe that was the cost.

C: But it happened between my first marriage and my second marriage, too.

BK: Well, that was the cost of that, too. I don't know.

C: So the cost as being?

BK: Having 30 women costs one marriage, every so many years.

C: But I would trade all of them …

BK: You're a man of plenty.

This is a major turning point in the flow of the session. His life will now be staged as "the man of plenty." In this act, we will discuss how the man of plenty will celebrate tonight with his woman friend.

C: I am a male.

BK: You are a man of plenty.

C: I am not.

BK: You are a man of plenty. You can come up with plenty of ideas, you can come up with plenty of relationships, you can come up with plenty of surprises, and everything I suggest you've already done or thought of.

C: Nice to meet you. (C shakes BK's hand.) (Laughter)

BK: Yeah. Quite interesting. So how are you going to celebrate tonight?

C: Gonna go to dinner. And I'm going to just play it by ear. Just gonna enjoy …

BK: That's all? Will you sing a song? Have you sung a song today?

C: No, I listen to the radio.

BK: Would you put on a special song, if you could find it?

C: No, I've avoided songs.

BK: If you could have a band … a live band tonight and have them play a song that marks the day as the end and the beginning, what would that song be?

C: I have no idea.

BK: Is that a song?

C: No. (C and T laugh.)

BK: We could make it a song.

T: We could.

BK: That would be good. (Singing) I have no idea whether this is the beginning or the end … (C laughs.)

T: That's a good song right there.

BK: (Singing and snapping fingers) No idea what it is …

C: Make sure you copyright …

BK: Frank Sinatra might sing a song like that.

C: I don't know. I have to be careful with music right now. Because it's so emotionally attaching sometimes. Because the other day a song came on the radio, an old Wayne Newton song, "Daddy Don't Walk So Fast," … I turned it off halfway through the song but it kept ringing in my head …

BK: So you need a new song.

C: I've been listening to the 70s station … and I have a couple of those CDs that are just like …

BK: See, here's the man of plenty. He has plenty of music. Everything I ask about, I find his abundance …

Everything is utilized to authenticate that this is a "man of plenty."

C: But I want someone to share it with. I want somebody to go with me. My career has taken off, like crazy.

BK: Get a dog. Less is more. (C sighs.) You're too successful with relationships.

C: I don't like dogs.

BK: Get a cat.

C: Oh, I hate cats!

BK: Get a mouse, get a pet mouse.

C: I used to have pet mouses. Black ones.

BK: See? … It's just unbelievable.

C: (Laughing) I'm sorry.

BK: Unbelievable. What was the name of your pet mouse?

Act V: The Mysterious Number "30"

C: I don't remember if I ever named them. Because they mated and there was 30 of them before I knew it.

BK: Same as the number of women. It's unbelievable. (C laughs.) How did you, how did the gods figure out how to put all of this in the world? I mean this is amazing.

I jump on the fact that what is plentiful in his life is quantified by the number "30." I am actually quite amazed by this, so I freely express my sense of wonder and surprise about how remarkable a fact it is that "30" women and "30" mice entered his life. We are in a new act, a world organized by the mysterious number "30."

C: (Laughing) Oh no!

BK: You have 30 mice as pets and 30 women who are solid relationships. That's unbelievable. This is getting weird. I'm starting to feel the room swirling around.

C: I want rights to this tape.

BK: Oh, man. Unbelievable. I mean this could be a television sitcom. This was the first premier issue of "Man of Plenty." We're going to call this show, "The Man of Plenty." Man this guy has it all. Doesn't know why he has it all. He just wants one little mouse as a pet and it turns into 30. He just wants one woman for a relationship, and it turns into 30. You know he asks for one song, and it's gotta be at least 30.

C: I just wanted one career and it's going like crazy.

BK: So what's the significance of the number 30?

Since we have established that "30" is a functional operative in his life, I ask him to articulate its significance. Perhaps this will enable the fulcrum to teeter toward an unexpected surprise ending where he has more creative presence in life than he ever imagined.

C: I have no idea what that comes from. Thirty was a bad year. When I turned 30, I cried.

BK: Really?

C: Uh-huh, uh-huh. I cried. Because I thought I had to grow up. I thought I had to grow up. I thought when I turned 30 years old, I thought, oh my God, I gotta grow up. And I cried. It was a terrible day, it was a terrible day for me.

BK: But you didn't grow up.

C: Uh-uh. 'Cause I called my mother that day. My mother goes, "Happy birthday," and I started crying. She goes, "Why are you crying?" I said, "I'm 30 years old; I gotta grow up." And she goes, "You'll never grow up, and you better not grow up."

BK: Your mom gave you permission not to grow up.

C: And I have not grown up.

BK: There we go. Now we got the bull's-eye.

The whole drama now centers on the moment in his life when the number "30" came into play.

C: I haven't grown up.

BK: Your mommy gave you permission to not grow up at 30. How many
    people get that? That's rare. But she not only said you don't have
    to, she said, don't.

C: (Sighing) So? I've made it through life.

BK: Interesting.

C: I have a successful career. I have a good job.

BK: I'm saying you have everything. The problem is that you have too
    much of everything.

At this stage, everything has reversed: the ending is a beginning, the
client is a therapist, the man who lost everything has everything, and his
problem is the cost of his accomplishments. The "Man of Plenty" is in a
creative topsy-turvy world where he stopped getting older at age 30. Now
we have a fascinating mystery.

T: Man of Plenty.

BK: The Man of Plenty. It's unbelievable.

C: Okay, I'm gonna go buy a cornucopia. Alright? (C and T laugh.)

BK: I'm getting an idea. How difficult would it be for you to let go of one
    of those 30?

I begin to plant the seed for a possible avenue of transformation.
Remember, he said he would trade all 30 for one long-term relationship.
Now we aim for dropping 30 relationships to 29, a small step toward his
desired goal.

C: I would trade all …

BK: It's easy for you to …

C: I would trade all 29 of them …

BK: No, no, we're talking about how difficult it would be for you to have
    29 instead of 30.

C: Well it wouldn't be …

BK: Who would you choose? Which one would go?

C: To get rid of?

BK: Uh-huh.

C: Um, probably, one of the drama queens. One of the ones that I just don't
    ever see myself with.

BK: I would guarantee you money the second he got rid of the drama
    queen, another one would just fill its slot instantly. That's what
    happens in your life.

C: Exactly.

BK: So you need 30 of everything.

C: (Laughing) Okay.

BK: The Man of Plenty, 30 …

C: I don't know why is it 30? Okay, let's keep …

BK: Initially, I would not have been attracted to this number. But when you said 30 women, and then you've mentioned that you had a pet mouse that turned into 30 pet mice, you can see why that would catch anyone's attention who was an astute observer of human experience.

C: Okay, I had rabbits, too. (Laughing) And I had 30 of them at one time. (T laughs.)

BK: Oh my. (BK stands up.) I gotta stand up, this is uh, this is too much. This is unbelievable. What's with this number 30?

C: I have no idea, but you got me thinking now.

BK: But he told us what it was with the number 30. Because on his 30th birthday, he was scared of becoming an adult and his mother gave him permission. His mother gave him permission to not be an adult.

C: Not to grow up.

BK: She not only gave him permission, she said he shouldn't grow up. So he's frozen 30 throughout his life. Everything comes in thirties.

C: You're right. (Pointing at BK) Because you wanna know something? This, this is really weird because …

BK: Whoa.

C: I think, you know how people always ask you how old do you feel?

BK: Yes. I know the answer. (BK stands up, and turns to T.) Let's go talk. Let's go talk. We need to talk about this. We'll be right back. (BK and T leave the room for a little over 2 minutes.) Extraordinary.

We have powerfully validated the mysterious presence of "30" in his life: he feels like he is 30, his mother told him to stop having birthdays when he was 30, he has 30 women relationships, and had 30 mice and 30 rabbits. We take a break to let him settle into this realization.

BK: You are one of the most interesting human beings I think I've met.

C: No, I'm not … I mean, come on. You've met hundreds of people. I can't be that interesting.

BK: You think of the details of what you've just reported. That your life got frozen at age 30, and 30 appears everywhere in your relationships with every living creature and thing. That is like an episode of *The Twilight Zone*.

C: And I don't like *The Twilight Zone* or *Star Trek* or *Star Wars* …

BK: No, you don't like the situation you're stuck in either. 'Cause you wouldn't be asking for something to change. So, yes, you wanted to be stuck at age 30 and never to go further, but at the same time, you don't like the cost of that.

C: Going forward ...

BK: Yeah, cause, it's ...

C: The cost of staying at 30.

BK: That's right, and you see ... you're probably going to ask for more therapy until you get 30 sessions.

C: How many sessions do I got?

T: Right now? This is 19.

C: Oh, well shoot, we got plenty of time! I'm sorry.

T: Wait, wait, what were you bargaining for, last week? What did I give you?

C: Two ...

T: We were talking about 10 to 11, was what you originally started with. 30.

Act VI: Move Forward by Experiencing the Missed Birthdays

BK: ... on the nose! Your whole mind is plugged into that 30 ... Do you want things to change? Do you really want things to change? In the good ways you're talking about?

Now a logic and plan for transformation will be introduced based on what he has expressed in the session.

C: Yeah.

BK: Okay, I'm going to suggest something.

C: Okay.

BK: The Man of Plenty is stuck at 30. We've got to loosen, free you from that number. Got to free you from that number. And this is how it's going to look ... How old are you?

C: Forty-five.

BK: The first day you feel like you've moved to becoming 31, you're going to feel that 30 change to 29. When you feel that you've become 32, you're gonna feel that number that's had a grip on your life become 28. As you move forward to the age that you are, that number is going to become smaller and smaller, until finally it catches up to being 1. I don't expect you to understand that, but I expect you to feel that it's true in some way, on some level in all of you. You know what I'm saying?

Because it was irrational that his life was stuck at age 30 and that the number began popping up everywhere, the rationale for the intervention is similarly irrational: as he catches up with his birthdays and becomes 45 years old, the tight hold the number 30 has on his life will be loosened.

What is important is that we are utilizing the same logic that is present in his life as a resource that can help him move on.

C: Okay.

BK: Don't think, don't even speak, I'm going to say it again. The Man of Plenty is frozen on the number 30—that was something that happened to you in your life. Be careful what you ask for. You asked for it, and permission was granted, and commanded to stay at 30. The number appears throughout your life, throughout your relationships. But, if you can move forward to feeling like you're 31, the number 30 will go to 29. And as you move forward each year, it will drop. So you have to find a way to catch up with the last 15 birthdays. I think he needs 15 birthdays as soon as he can tolerate them. And for each birthday, to grow 1 year. I don't know how you're going to do that, exactly. Maybe tonight should be a birthday party. Tonight you should celebrate becoming 31. Can you do that?

C: Yeah, but I don't understand this, what I don't understand …

BK: I don't understand why you have 30 rabbits, 30 mice, 30 relationships, and at age 30 your mother said you have permission not to go further. I just know the 30 stands there loud and clearly for the Man of Plenty.

C: Oh my gosh …

He has an "aha" moment: "30" is organizing his life and it began when he decided to not be any older than 30. Understanding or explaining why this took place is less relevant than recognizing that it accurately describes his reality.

BK: And that if you can begin to have birthdays move forward year by year in the development of who you are, then with that will be a corresponding decrease. Because it will be a loosening of the grip of the age 30 on your life.

C: Okay.

BK: I'm not asking you to understand it; I don't think it's understandable. I think you're a big mystery. But the mystery holds these connections and holds these facts—and that we can work with. … Tonight go celebrate your birthday. This means the friend you're with, or you, must ask the waiters to sing "Happy Birthday." Go to a place where they sing "Happy Birthday."

C: Oh, that would be so embarrassing. I couldn't do that to myself. Oh my God, I couldn't do that to myself.

BK: See?

C: I might sing "Happy Birthday" to myself.

BK: You could sing with them. I'm not asking you about embarrassing yourself. I'm asking you to celebrate your birthday tonight so that you feel that you're 31.

C: I have not had a meaningful birthday …

BK: Of course.

C: Since 30.

BK: I mean, look, "since 30." (Pause) There it is. Do you realize … (BK stands up and walks over to C.) Do you realize what's happened to you today? You found what you're looking for.

C: It's like an epiphany.

The context that holds and gives meaning to his life has been transformed. He, the Man of Plenty, has been frozen at 30. Now it is time for him to seriously entertain action, moving on with his birthdays.

BK: You found the answer. It's all showering down. It's pouring on you. … Everything that comes is verifying, confirming this. I'm gonna insist you go buy yourself a birthday card, too. Get a birthday card for yourself, and buy yourself a birthday present. … Have them bring out the dessert. … If they sing it, it's fine. You can just tell your friend that your therapist insisted you do this. (C laughs.) You could even make something up and say, you know, it's really the first day of the next day of my new life.

C: That's what I was … thinking. I've been playing this out. You know, this is the first day …

BK: Will you do this? (BK steps back.)

(C sighs.)

BK: Will you do it? Will you take a step forward?

C: I will, I will, so heartily, strongly try …

BK: Nah, that's not good enough. (BK sits down.) No. I don't buy that. No. I don't want to leave this room until I hear you tell me that you will do it. I'm asking you to have a birthday tonight. I need to hear that. Otherwise I'm just going to assume that number 30 will be present for you the rest of your life. It's now. Let me ask a different way. Which of these things will you agree to? Will you agree to get a birthday card?

I am helping bring forth all the small actions he can agree to that define having a birthday. Block by block, we will see if we can construct a birthday party.

C: Sure.

BK: Okay. Promise?

C: Promise. (BK and C shake hands.)

BK: Birthday card. Excellent. A small present?

C: Sure. (BK and C shake hands.)

BK: Okay. At the meal, request a birthday dessert? (C sighs.) You have a choice. Or you bring a dessert, with a candle.

C: I could buy a cupcake and put a candle in it.

BK: You'll do that?

C: Yes, sir.

BK: You promise?

C: Promise.

BK: I'll give you the option … (BK and C shake hands.) … will you consider that to be a birthday?

C: Sure.

BK: And will you then tell her whatever will make sense, so you'll feel comfortable in describing this as a birthday that is the first day of your new, next life … putting this into the social realm is very important. I don't want it to be private. 'Cause a real birthday means somebody is with you on your birthday.

C: I think here's my dilemma, I think if it was anybody else …

BK: Then she's perfect.

C: … I would be okay.

BK: Anybody else would be too easy. Then this is perfect. For it to be a real birthday …

C: 'Cause this is the first time her and I have ever gone out …

BK: Oh, fascinating. So you could you use that … she knows you had your divorce today, right?

C: I don't know if she knows or not.

BK: Hmmm … Hmmm … Very interesting how the universe aligned this up so that if you're going to do it, it's really going to be a big bang.

C: Okay, okay …

BK: Would you …

C: I will order dessert, and I'll get a candle, and I'll put it in it, and I'll sing "Happy Birthday" to myself.

BK: Perfect.

C: Embarrass the crap out of myself. Why not? (BK and C shake hands.) Done worse.

BK: Unbelievable. So, um, I'm very grateful that I was able to meet you. Because you truly are a remarkable human being … Yes, it's a given that you have skills with relationships. People call you, people are attracted to you, on every level. Wanting help, wanting mothering, wanting fathering, wanting relationship, including long-term relationship. You truly have gifts. Both as a nurse, as a friend, as a person some want to enjoy, and as a therapist.

But wow: Did you ever rub Aladdin's lamp, and get what you asked for!

I put a positive connotation on his life: He's been too lucky rather than unlucky. He received what he asked for when he made his wish.

C: But I don't feel that way.

BK: Doesn't matter how you feel, you asked for it. The Man of Plenty asked, and mom gave you permission, and it got stuck at 30, you haven't had a birthday since 30, and everything in your life that involves relationships is around the number 30. I don't want you to think about it anymore. You're not going to be able to understand it. I just want you to concentrate on one thing: catching up with those birthdays. Tonight's the first one. Maybe the next one oughta be, I don't know, next week, next month, I'm not sure. That'd be … a good reason to have a next session. Have one more session, where you focus, the two of you, on how you're going to catch up with the other birthdays. Okay?

T: So should we do one more session, and then our own birthday party kind of thing, or just one more session?

BK: You'll figure it out. You guys can figure it out, okay? That's enough, I don't think I can take anymore. If I hear another 30, I might, I might …

C: Well see, now I'm thinking, now my mind is like, okay now my mind is rewinding my life okay?

The context that holds his life has moved into a new home. Now he is experiencing how the experiential belongings of his life fit into the new contextual home he has moved into.

BK: Oh no. It's gonna be a lot, you're going to find things that are gonna just surprise you. But don't worry about it, don't try to understand it, just know something, something miraculous happened to you, which means that you're able to attract things in the universe besides people and bunny rabbits and mice. You attract things and its, uh, there's some gift you have and, and this is in itself good, you just need to be released so the rest of the gift can come forth. And other blessings will come from you to others and come to you for you. Okay?

C: Okay. Thank you, sir. (BK and C stand and shake hands.)

BK: Yes.

C: Nice to meet you.

BK: Nice to meet you. Happy Birthday.

C: Thank you, sir.

BK: Okay, that's it.

T: Cool.

BK: You guys schedule something for next time. (BK exits the room.)

T: (C and T stand up and start walking out of the room.) Happy birthday.

C: Thank you.

When scheduling his appointment for his next and final session, the client told the student therapist, "I can't stop thinking about this. I keep thinking about all the 30s in my life. I can't believe this guy changed my life in 30 minutes."

Four months later, I received a letter from one of the faculty members of the university where this session had been conducted. It read:

> Brad,
>
> I hope all is well with you. I thought I might drop you a note about a chance encounter my dad had with the "Man of Plenty." My dad [who is director of the clinical program] was unfortunately in the hospital because of some blood clots that were found in his leg. He is doing well now ... However, his nurse for a couple of nights was the "Man of Plenty." He recognized my dad and proceeded to tell him how much the session he had with you changed his life. He updated him on his aging process that you suggested and his plan to be at his current age at the time of his next birthday ... I thought you might like to hear that he is doing well.
>
> Take care,
>
> Justin Moore, Ph.D.

A year later, the "Man of Plenty" faced a major life catastrophe and was uncertain about his future. I returned to the university as a visiting professor and held another session with him. This time, he had read many of the books I had written (at that time, I had published "30" books). He was also arguably at his lowest bottom. Without missing a beat, we immediately stepped into the reality of his being a "Man of Plenty." Along with the many positive things in his life, I pointed out that he had also received plenty of hardship and suffering. It was up to him to utilize this to make his life richer, more complex, and potentially more transformative. I invited him to embrace the suffering that came to him and convert it to being another source of learning, uniqueness, and hard-earned wisdom that could be used to help others in ways he never had expected before. He left the session feeling more plentiful and ready to make a contribution to life than he had the first time we met.

CHAPTER **5**

A Night in the Love Corral

A married couple, in their mid-sixties, came to therapy with a presenting complaint about the husband secretly buying horses without his wife's consent. The following interview was a second session conducted by a student trainee (T) at a university family therapy clinic in Arkansas. The husband (H) initially discussed his recent business travel to a horse show. He makes his living selling blankets, bits, saddles, neck sweats, and other horse wear.

The student therapist asked the couple if they had done their homework assignment to make a list of all the things they had enjoyed about their marriage over its 22-year-long duration. The wife (W) pulled out a sheet of paper with her list, but the husband had not followed through. She mentioned that one of their marital highlights was a trip they took to Yellowstone Park years ago. She initially had dreaded the trip, thinking that they wouldn't be able to keep a conversation going for that long. She even took audiotapes to keep from getting bored, but it ended up being a good time.

Act I: Olympic Silver or Gold?

(Knock on door; Brad Keeney [BK] enters.)

T: Come on in …

BK: Nice to meet you. … I want to ask you a few questions. Do you consider yourselves perfectionists? (Looking at W) Are you a perfectionist, do you ascribe towards perfection? Would that be a fair description?

107

I ask this given my observation of how neatly she was dressed and the manner in which she presented herself. In addition, she had her homework neatly folded and tucked away in her purse and had faithfully followed the assigned homework.

W: Probably, somewhat.

BK: That's an accurate representation of the way in which you participate in life, to seek perfection? (Looking at H) And parts of your life, would this also be true? Or some part in your life—that you seek perfection?

H: Naw, I'm not a perfectionist.

BK: So she seeks perfection. How would, what metaphor, what word would describe the way you are in life?

H: If it works well, that's okay.

BK: Okay.

H: It just needs to work well. Yeah.

BK: So if it works …

H: Don't fix it.

BK: Don't fix it. (To W) For you, if it works, make it better.

W: (Nods) I guess so. Yeah.

BK: So my question to both of you now, with respect to your marriage, do you want to go for the Olympic gold or the silver? Which medal do you want?

This is a classic illusion of choice because both outcomes are victories.

H: Silver's fine with me.

W: At this point, silver looks good to me.

BK: But you'd rather have gold.

W: Well, yes, but I mean …

BK: So she wants gold, but you'll satisfy yourself with silver.

W: That's right.

BK: (To H) You're okay with silver, but you'd be okay with gold.

H: Oh yeah.

BK: I mean if you could arrange it, to have a gold just to make her happier, even though you didn't care, would that be something that you could accept?

H: Sure.

BK: I heard these lovely reports of your traveling, and you clearly both have a life that's good. I can tell by the way you sit in the chairs that there's a sense of pride about who you are and how you are in life. … It's how you sit. And there's a professional manner with each other. You're so clearly not being taken to the ground by problems and suffering and hardships that are so horrific that you have to crawl through the door and barely be able to say a word without overwhelming yourself with tears and sorrow and pain and anguish. You're coming in here, two people … thinking about going places and … either a silver or a gold we're gonna go for. Now, how much time do you have with each other during the course of the week? 'Cause I heard that you're off and about. How much time do you have in each week?

I am making positive connotations and moving on to see how much time they have to work for the medal.

H: Probably 3 days a week.
BK: So you've got 3 days?
W: Not together, though. That, even if …
H: (Interrupting) No.
W: That's one thing he fussed at me last night about, because I've been spending too much time in the school, especially while he's home. We haven't had much time together.
BK: He's on the road a lot, and you're at school a lot, and then the intersection might be at best 3 days where two bodies occupy the same physical proximity.
W: Right.
BK: And in that 3-day period, how much time in a week do the two of you actually have being with each other?
H: Well, she got home what, 7:00, 7:30 last night. She went to bed at 10:30. That's about the routine.
W: But ask him what he had been doing while I was home.
H: Yeah.
W: He was out riding his horse.
BK: Okay, so that intersection when you're in the same area is 3 days, but being in the same place, at the same time, focused together, with a sense of being together, is very rare.
W: Very rare.
BK: (Looking at T) So, this is a team, going for the Olympic silver or gold, that's not even getting on the basketball court.

The metaphor of Olympic medals arguably enables us to talk about their issues in a softer, less resistant manner.

BK: What do you do in those 2 hours [that you do have together]?
H: Well, we don't sit and talk very much. That was a conversation we had yesterday. I was not happy … (mumble)
BK: You're both a little creative, too, aren't you?
W: I don't look at myself as being creative; he is.
H: Yeah. I'm a creator.

Act II: Perfectionist Wife and Creative Husband Make Plans

BK: Okay, she's a perfectionist; you're creative. That's a winning combination. If we can get perfectionism with creativity together, we're going to have a gold medal. We're going to have a creative silver medal, which is gonna be a perfect gold medal. Okay? So, I think it'd be interesting for us to think about the ways in which you could creatively do things together. What could you do together, that you've never even imagined. I mean, where do you eat your dinner? You live on a farm? Is that correct?

Now that we have a metaphor for each spouse—a perfectionist and a creator—we can move onto a practice field, the next act of this theatre of creative transformation. Now we need to find a field.

H: Correct.

BK: You ever have dinner in a barn?

H: Yeah, we've had dinner in the barn with college kids. That I do.

BK: How about just the two of you?

H: No.

W: We have. (Talking simultaneously with H) We've had wiener roasts out in the barn …

H: Well, we have had wiener roasts …

BK: But you know what I mean. I'm talking about a date in the barn.

W: I mean, I go out and help him feed, but that's not …

BK: I'm talking about the two of you.

W: That's still not …

BK: You know what I'm talking about.

H: I know what you're talking about.

BK: So, what would that look like? Who would prepare that? Who would set that up?

H: Me.

BK: You'd set that up? You'd put up a table, maybe put up some candle light?

H: A good one.

BK: Set the meal up.

H: I [can] do that. I'm a good cook.

BK: Would that make you happy?

W: Sure.

BK: Boy, that was easy. (T and W laugh.) That was easy. We can just straighten up about 12 of these things, you all will be halfway there. All right. So when can you do that?

With a creator and perfectionist on a playing field, the session will now move along on its own, being pulled forth by the creative ideas that come forth effortlessly.

H: I leave in the morning.

BK: You leave in the morning? You coming back? When you coming back?

H: It'll probably be Monday.

Act III: Horsing Around in the Love Corral

BK: Okay. You ever spend the night in a barn?

H: In the park?

BK: Besides having dinner, have you ever spent the night in the barn?

H: Yeah, I have.

BK: Both of you together?

H: No, just when the mares have foals.

BK: Now that's another interesting idea. Not only have dinner in the barn, you might even spend the night in the barn. I think you could set that up, too.

H: Yeah. Yeah.

BK: Be very interesting. Would you bring music out there?

H: Yeah. I like music. Some kind of music.

BK: Do you know the kind of music she likes?

H: No.

BK: You don't?

H: Huh-uh …

BK: Well, let's find that out right now … (To W) What's, what's music that charms you?

W: Some of the older music I guess … I like country music. That older music.

BK: Can you tell him a song?

W: Gospel music.

BK: Can you tell him a song, or an album, or an artist? That if you had the night to remember, the night to remember in our Arkansas barn, and it had music, music that made your heart warm … made you …

W: I could probably pick one out of a list, but as far as coming up with one right offhand …

BK: What about Tennessee Ernie Ford?

W: That would be all right.

BK: You like Tennessee Ernie Ford?

W: Yeah, or Johnny Cash or …

BK: Johnny Cash.

W: … Elvis Presley or …

BK: There you go. You should have all three of those on hand.

H: I don't have Johnny Cash, but …

BK: You certainly can get a hold of Tennessee Ernie Ford and Elvis Presley.

H: I got plenty of CDs in the truck.

BK: Really?

(W nods.)

H: Pete Fountain.

BK: Really? Dixieland band. I love Dixieland, too. (Looking at W) You like Dixieland? (W nods.) You might progress it. You might start with Tennessee Ernie Ford … I'm not sure were Elvis goes. Elvis is pretty hot.

H: Elvis has got some good songs.

BK: Maybe an Elvis ballad. Then move, when it gets to dessert, to being something like Dixieland. Where the music gets a little wild. … Well let's think of a three-course dinner. First course is going to be the Tennessee Ernie Ford. What kind of food would go with that? Maybe that'll be your starter.

Here the use of therapeutic metaphor is clear: stages of romantic interaction that are isomorphic to the sequenced courses of a dinner. They are no longer in therapy. They are playing around in the barn.

H: Tennessee Ernie Ford is like barbeque. I do that a lot.

BK: And if you had another course following your barbeque, that would be Elvis, and what would go with Elvis? Don't say peanut butter. (Laughter) I read that cookbook.

H: With Elvis, I would say a little more spicy kind of thing you know.

BK: Excellent. Excellent. (To W) Does this sound good to you? And then for the final course, when the music is really wild, Dixieland, what food would go with that?

H: Oh, that would probably be like shrimp and … it's been a long time since I've done that, but I created a shish kebab with shrimp and peppers and onions and everything, and added it all together.

A shish kebab, or stick of grilled meat, is an interesting choice of metaphor. He's pointing the way for how we should talk about their barn party.

BK: (To W) Have you ever had that?

H: Yeah, she's had that.

BK: So that's been in [both] your time. That wasn't when you were a boy. You've done that …

H: We've done that at the house.

BK: I really want to see the two of you make a commitment to doing this. I think you're going to be in for a big surprise. I don't want you to hold back. I want you to use every drop of creativity you have in your being to pull this off, to make this night [one that] you'll both remember. It should be three courses. The Tennessee Ernie Ford course with the barbeque, followed by the Elvis course with the spicy, and followed by the fireworks of hot Dixieland, and wild music, and the most spectacular variety of assortment you can put together on your shish kebab. Okay?

BK: (To T) I'd just like for you to … give some further consideration about this, but I think this night in the barn, this three-course dinner, is going to be something. I think you'll both like that idea, don't you?

(BK leaves the room so that the therapist and couple can talk further about the idea of having a night in the barn. They try to think of a date when it might work and the preparations that would be necessary.)

As in the previous case, I leave the room in order to let the clients settle into their new contextual home—a night in the barn. When I return, we will continue elaborating the resourceful context, adding more metaphor and activity.

(BK knocks on door and then enters.)

T: Come in.

H: My barn is not concrete or anything; it's just dirt.

BK: Good, it's more romantic.

W: We just thought it'd be a little more difficult to spend the night out there.

BK: Well, your husband's a creative guy.

BK: What I'm really happy about is that you're now in the arena. You're on the playing field; you're on the court. Now let's add some things to help you get the gold, okay? … I want you to consider how you're going to come dressed. And since there are three courses, there should be three things associated with the way in which you dress. 'Cause the way we've structured it, the first course with Tennessee Ernie Ford and the barbeque is gonna be something more formal … You ever have white gloves?

(W nods and laughs.)

BK: You got some white gloves? I'd recommend …

White gloves could be taken as a metaphor for her perfectionism, but more importantly, they are a clothing item designed for removal.

W: I have had, I don't know if I could still …

BK: See if you can find them. … There's something you could wear for the first course. Then take [the gloves] off for the second course to be more informal. Right? … For the second course, you're gonna be looser. But on the third course, it's gotta be a surprise. You know? My, my, I can't even begin to imagine, I don't want to give him any ideas. It could be that you're wearing a color of stocking or something that you haven't revealed or that you have a pendant and that [when] you just turn around, do something that surprises him. But it would involve the manner of what you would wear, or not wear. In this regard, each stage will be a shift in how you present yourself. Okay? Okay, that's very important. (To H) When's the last time you shot fireworks?

H: I haven't shot fireworks in 20 years, maybe 30 years.

BK: I'd like to see you have some fireworks.

H: My neighbor has fireworks stands. We got all the fireworks at our house.

He has all that he needs. But why stop now? We keep on with the night party that is so graphically described that their imaginations are already inside the barn during the session.

BK: Fantastic, it's perfect.

T: Yeah.

BK: Perfect. When this whole extravaganza, when this whole evening once-in-a-lifetime event in your barn takes place, I'd like you to arrange a fireworks show for your lady.

(Nodding all around.)

BK: And I don't want you to stop thinking about it … Every little detail that you might contribute, might add. Use your creativity, use your perfectionism. (To W) You might even … do you like to write things? Are you a writer?

Now I will bring her more into the scene, utilizing her talents.

W: Not really.

BK: Or someone who writes things, thinks things clearly?

W: I don't write poetry or anything like that.

BK: No, but you think clearly, right?

W: Well, sometimes.

BK: You're a thinker. You're a person with words, yes? I don't mean poetry, but I mean, if you need to express yourself, you're good at it. You might even have something that you pull out and read that's a special thought. He's bringing the music, you bring the lyrics. You're going to have the music, you bring some words. For each course, you can read something. What kind of words would fit Tennessee Ernie Ford—for being a little more formal? What kind of words would then fit Elvis? What kind of words would fit hot Dixieland wild music? You're going to be mindful of how you dress. What you reveal, what you conceal, what you show, how it moves from formal to informal, and words that accompany that. (To H) You're preparing the meal, you're bringing the music. And at the very end, you're going to celebrate this whole special night with fireworks. You ever sing? Are you a singer?

H: Me?

BK: What about you singing?

W: He can, he can.

BK: You're a singer.

H: I sing to myself.

BK: Have you ever sung to your lady?

H: No.

BK: Well we're going to change that, aren't we? Gotta go for gold: you've gotta go with everything you've got. See that's the only rule here. The only rule to this, if you're going for the gold, [is that] you've got to use everything you have. Maybe when everything is over, and the fireworks have been shot, and you're standing outside under the sky, you sing a song. You sing to her. It'll come forth in a way that's different when you're standing outside in the dark. Sometimes you can say things to each other more easily when it's darker. You know?

"Going for the gold" means using every resource, gift, skill, and talent that they each have for one another. Doing all of this in the dark—the time of romance and mystery—is easier than doing it in the brightly lit room of a therapy session.

H: Yeah.

Act IV: Being Married

BK: It's a lot easier to express your heart when it's dark. When it's bright like this, it's hard, but if we make it dark … (BK gets up and turns off the lights in the room.) … it's a lot easier to talk. Isn't that strange? I mean it's funny, when we can't see, we sometimes feel like we're closer to each other. You feel that? I better turn the lights on.

W: Where are you H? (Laughter and the wife is seen having reached for her husband's hand.)

I turn the lights off in the room so that they may fully step into their imagination—the dark make-believe world of fantasy located outside the margin of everyday habitual seeing and knowing. When I turn the light on, we notice that she has reached for her husband, a moving moment that I immediately highlight.

BK: See, she was reaching out for you right there. See there? I'm excited about this, 'cause I really think you are a creative man. I think you're a woman who seeks perfection. And my, oh my, what we're looking at right here between these two people is a marriage between perfectionism and creativity. They've not quite fully found how to be with one another, and when they do they'll be a new marriage. They're already married in name, but the creativity hasn't yet been married to the perfectionism. The perfectionism hasn't yet been married to the creativity, and when that happens, whooh! You're gonna have gold, you're gonna have barbeque, you're gonna have spicy food, you're gonna have surprising shish kebab, and you're gonna have every kind of music and fireworks. And that barn is gonna have to be renamed. I don't know what you're gonna call that barn. Maybe you oughta put a sign up. Put a sign over the barn door—maybe it's "The Love Corral." I don't know. But put a sign that renames this for the special night. You're feeling this with me, aren't you? I'm feeling it. I mean, this might be so good [that] we make a TV movie out of it.

The gold medal winning marriage is defined as a union of perfectionism and creativity, something that can be imagined in a barn party.

H: Yeah, that'd be great.

BK: That'd be great, wouldn't it? Okay? It's been a pleasure. I can't wait to hear about this. Will you tell me?

T: Oh yeah.

BK: Great. Alright. Real pleasure. (Leans over to shake H's hand.)

H: Nice to meet you.

BK: Nice to meet you, creativity. (Leans over to shake W's hand.) Nice to meet you, perfectionism. (Grabs each of their hands, and places their hands together.) I pronounce the two of you man and wife. (Laughter.) Okay. (BK exits the room.)

We enact a brief marriage ceremony, having them hold hands while pronouncing creativity and perfectionism as husband and wife. I depart so the therapist can keep the newlyweds in the barn for a little while longer.

T: Good. So, you both have your roles, you know what you're supposed, what you're planning on doing. I mean, obviously you don't have it planned yet, but you know you're gonna be …

W: What I need to do.

T: You know what you need to do. Okay, so you're in charge of the words, the lyrics, and the dress, the dressing. The presentation.

W: Maybe I ought to write that down.

T: You wanna write that down?

W: Words, lyrics, presentation. Me.

T: You, yes. And you H, you know what you're gonna be doing?

H: Uh-huh

T: Gotcha. Doing food, fireworks, music.

H: Uh-huh.

T: I am excited … to hear about it because you are creative. I mean, just thinking that, what you've told me about, the stuff you've done in the past with jobs. You know? You've thought up some awesome stuff. But I think that this is gonna be a wonderful time to be able to dazzle her with your creativity. You never known that you could woo someone with your creativity?

H: Do what?

T: You could woo someone with your creativity.

H: No, I hadn't thought of that.

T: It's that attractive. Or with your perfectionism. Someone could delight in your perfectionism. That's what a marriage of creativity and perfectionism is.

(BK knocks on door.)

T: Come on in.

BK: (Standing in the doorway) Am I right that your buying horses is the reason that got you into this counseling?

H: No, that's just one of the things.

BK: But you see, you see how creative he was? That's what brought you into the barn. Man, that's creativity. (BK exits.)

Even the presenting complaint is now a resource when it is fenced in and experienced inside a barn party. The newlyweds have now, in fantasy, horsed around in the love coral. Now they walk away with the most resourceful parts of themselves joined together in Olympian matrimony.

The couple subsequently reported no more need to have therapy. We'll never know what happened in that barn, nor should we know. This session gave them a new marriage: a wedding between her perfectionism and his creativity. As their imagination went to the love corral in our session, new possibilities for how to enjoy one another were awakened. Finally, those horses he secretly bought were never a problem, though they seemed that way before therapy began. After this session, the horses could be seen as simply leading both of them into a love corral and into the barn of their imagination, where a marriage can find all kinds of surprises and fireworks!

CHAPTER 6

Weight-ing to Leave

A family consisting of a mother (M), father (F), 18-year-old teenage son (S), and younger 12-year-old daughter (D) had been receiving treatment at a university family therapy clinic. The presenting problems concerned their adolescent son, Bob, and were initially described as his poor grade performance, obesity, and difficulty getting a girlfriend. In this session, which I am initially observing from behind a one-way mirror, the student therapist (T) has been following up about the boy's grades. His algebra grade successfully rose from previous test scores in the 40s to this week's scores in the mid-90s, enabling the mother to boast that the therapist was smart in helping them have a game plan to improve the boy's performance that worked. The father quickly added that their son must still pass the semester exam before he can graduate from high school.

Act I: The Son's Problems

S: I could probably fail it and still go on.
M: Let's not test it.
S: But, yes, I want to pass it …

The son's dramatic improvement in algebra indicates that he has the intelligence to do well in school. However, in response to his father's concerns about the semester exams, the son shows a hint of ambiguity about whether he will pass them. This exchange signals that his school performance may be organizing family dynamics.

T: I think he has come a long way. Right?

(F laughs.)

S: Apparently, Dad has a different opinion because he's laughing.

F: No. You have in some aspects …

T: Right now he is moving forward to graduate from school.

F: Uh-hum.

T: He's also ready to move forward on his weight issue and ready to move forward with getting a girlfriend …

F: Fishing in a different pond, right?

S: (To parents) Now tell me. Have I improved in my weight at all?

F: I told you the other morning you have.

S: (Looking at Mother) I want to hear it from you.

M: Improved in your weight? I don't know. Have you lost any in the past week or so?

S: See, that's what I meant.

M: Baby, I can't tell. You haven't told me anything.

S: I shouldn't have to tell you anything if I lose weight. You should be able to tell.

As I observe this interaction from behind a one-way mirror, I am aware that the son's focus on who is noticing changes in his weight (or life) is more important than whether he has accomplished an individual goal. Clearly, the family is showing us that the so-called problem is a metaphor they use to dance with in their interactions with one another.

T: I think you mentioned that you lost four pounds in two weeks. You said four. Have you lost some more?

S: I don't know; our weight scale is out of power.

M: I don't mean to be insensitive about it, but the bigger you are …

S: You're insulting me.

M: I'm not trying to insult you. I'm big. The bigger you are, four pounds doesn't show as well. You know, on a thin person, four pounds would show quicker than four pounds on a bigger person. That's why I haven't noticed it.

T: Why do you want your mom to know about it since you know that we have a game plan that she shouldn't have to talk about it?

S: She should just notice. I lose weight. (Looking at Mother) You should notice; that's just how it is.

Again, I observe that weight is a surface concern; it is a metaphor they use to discuss how they interact with one another.

F: Well, I know you've been eating different. You're eating Special K. … He eats a bowl of that. When I'm laying there in the morning, he's eating a bowl of cereal.

S: That's what I eat for breakfast and supper!

T: My understanding is that you're not thinking about graduating from high school at all, but about graduating from the various problems or the various goals that you have set for yourself. Right?

S: The weight is probably the biggest one.

F: How much did you lose last summer?

S: Twenty-five.

F: You'll do it this summer. It's going to be hot and you'll be outside working.

T: You said you wanted to lose about 50.

S: Uh-hum.

T: And you've lost about …

S: Not much.

T: You didn't buy into what I offered. Do you remember what I offered?

The therapist and family discuss how the therapist had previously teased him to gain more weight because maybe he would reach a weight where things would start to self-correct.

(Brad Keeney [BK] enters the room and shakes hands with everyone.)

BK: Hi there. … Is weight the problem here? Is that the main thing we're working on?

T: It's one of the problems.

BK: Is that the main one?

T: No. He has been trying to get a girlfriend and every attempt fails.

BK: Okay.

T: And he thinks that the weight is probably part of the problem.

BK: Are you trying to lose 50 total?

S: Uh-hum.

S: Where in there were you? (pointing to the one-way mirror)

(F laughs.)

BK: You could tell?

S: Yeah.

BK: Do you have X-ray vision?

This exchange actually sets in motion the idea that the son sees something that others do not see, or voice, or express.

S: No, it's just too obvious.

Act II: Weight as a Metaphor of Family Interaction

BK: I heard a little bit of your conversation. (Pause) Now this is a crazy
   question.

S: What?

BK: I've been at this kind of work for a long time, and I have found that
   sometimes a question you never expected being asked might just
   open something up. (Pause) Do you think that the rest of the
   family would care enough about your losing weight to promise
   you that for every pound you lost, they would gain that pound?

I first announce a shift toward a possible new direction and then ask a
question that brings forth the relational interactions that hold his weight
control. This moves us away from the first act, where weight was the adolescent's individual problem, to the second act, where we are more explicit
about weight being a way of discussing family interaction.

S: They care that much?

BK: Uh-hum.

S: I don't know. I wouldn't want them to gain it.

BK: Oh no, I'm not talking about being logical or rational. (M laughs.)
   But, just if …

S: Yes, yes, I do … I do honestly.

BK: Is it true? Would you gain 50 pounds if it meant he lost it?

F: Yeah.

(M nods her head yes.)

S: (Pointing toward M) Ask her that question.

F: Well … I know the answer to that. (Laughs)

BK: Would she do it for you?

S: No, she wouldn't.

BK: She's smart.

S: Yeah, she is.

BK: She's smart. She'd take care of herself. She knows how to take care of
   herself, doesn't she?

S: Yeah, except for eating candy all of the time.

BK: Does she know how to take care of herself better than you know how
   to take care of yourself?

(S shakes his head no.)

BK: What do you guys think?

M: No.

BK: Are they about the same?

M: Yeah.

BK: That's good. I'm just checking; just testing. (Pause)

This would be interesting to see because it's just as much work to gain as it is to lose. I don't want them to gain 50 pounds, but it would be interesting if they gained just a couple of pounds … just to see if he could see that they gained.

F: (Laughing) He ought to be able to see that I've gained here in the past month.

(M laughs.)

S: I have.

BK: So you (the son) have better observation than he (the father)?

S: Uh-hum.

BK: Is that true?

F: No, 'cause I've noticed that he's …

S: He has noticed.

M: I haven't.

S: My Dad said he noticed and my Mom has not said that she noticed. And she's the one that has lost a lot more weight than anybody that I know.

BK: Does she have the most skill at losing weight?

S: Apparently.

BK: Who is the best observer? Is it you two men?

S: The woman. The woman over there is supposed to be. (Looks at M.)

BK: That's the observer?

S: Supposed to be, yes.

BK: But she didn't notice.

S: No, she didn't notice.

BK: So, that's confusing.

We are talking about a mother who should notice change (of weight) but doesn't and men who notice change but aren't as good at noticing it as mother. The family issue is less one of weight than of who is noticing change. What is about to change? Why the concern with change? Why would weight be the metaphor for change? We shall see what is brought forth.

Act III: Bob's World

S: (Shaking head) Welcome to my world.

BK: So, it's confusing. It's confusing to be in your family?

S: Confusing to be in my world.

BK: It's confusing to be in your world and your world includes your family?

S: Yes.

BK: And girls and school ...

S: Yes.

BK: If you could go to any world in the world, where would that be?

S: My world.

BK: Where's that? Is that a name?

S: Bob's world.

BK: Bob's world.

S: Yeah.

BK: Yeah, that might be a TV reality show. (M and S laugh.)

S: I love my family.

BK: Yeah?

S: They don't think I do, but I respect them.

BK: Why don't they think so?

S: Because I don't act it. Now I could understand that because I let my frustration be known very easily.

BK: Um-huh.

S: And I smart off to them very easily.

BK: Yeah, how old are you?

S: I am 18.

Act IV: Leaving Home

BK: That's what 18-year-olds do. (Pause) By the way, when he leaves home, he's going to break your heart.

I am providing an entry to the change no one is talking about: Bob's graduating from school and the possibility that he may leave home. As a family therapist, I am assuming that metaphorical talk about noticed and unnoticed change includes how the family is preparing for the big change associated with Bob's departure. The family is facing the possibility of losing a lot of family weight—Bob's presence at home. Now that we have shifted to this thematic context, conversation can come forth about how they are handling this developmental shift in their home life.

F: Yeah.

BK: So Bob sort of has to act like this ... He has to act a little obnoxious because it makes it a little easier for him to leave.

In this thematic context, I am in a position to mark all communications they present as being about how the family is dealing with Bob's leaving home.

F: He's in on a, on a …

M: I'm sure he's doing it for that reason. (Laughs)

F: Learning curve. He wants to start at the local college next fall and live at the dorm. I've done told him he's not bringing his dirty clothes home for mama to do.

S: I already know that. There's no reason to tell …

F: They've got a laundromat here and all of that so he's going to be in for, on a learning curve.

M: We want to have a good relationship.

BK: I know that. I know that you know that.

S: Yeah.

BK: Yeah.

S: My problem is the whole thing. I love them to death, but they tell me things that I already know.

BK: Parents say that because part of them wants you to bring the dirty laundry home so they can see you more often. They don't even know that.

The "confusion" that Bob previously indicated as characterizing "his world" now makes sense—his parents both want him to stay because they don't want the experience of missing him and they want him to move out and get on with his becoming a successful adult.

M: Um-hum.

BK: See it's going to be …

M: Shhhh.

BK: I know. There are going to be all kinds of mixed communications because they both want and don't want this to change because leaving is a hard thing to do.

T: Um-hum.

BK: I bet they wish that you had gained 500 pounds because then you wouldn't be able to move and you would have to stay home. But they don't really want that …

In Act IV, "Leaving Home," we are able to understand how weight is a metaphor for leaving home. It also helps us appreciate the contradictory messages that have taken place about it. Mom, the expert observer, doesn't want to notice any change. She is arguably struggling with her son's leaving home.

S: Ah-ha!

BK: You know what I'm saying?

S: Yeah.

BK: Parents don't want their kids to leave home. That's just what love does to you. It's hard … when it starts to be time to leave home, all kinds of weird stuff starts happening. This is true for you, too. Part of you, as much as you want to get out, or at least you think that you want to get out ... but there's some advantages to not leaving ...

S: There are advantages.

BK: ... of being home.

The theme of leaving home brings forth the confusion—the contradictions of wanting to both stay and leave—that everyone feels.

M: I told him that I didn't think he ought to get a girlfriend. He shouldn't worry about a girlfriend because ...

BK: What does a girlfriend mean? It means he's leaving home.

M: It means no college.

BK: It means he's leaving home. (M laughs.) It also means he's left the family.

M: Yeah.

BK: If you get a real girlfriend ... he doesn't quite want to leave the family and you don't want him to quite leave the family, so you are all invested with all the things that are happening. You see.

All of the original presenting problems can now be seen as metaphors for leaving home. Weight, girlfriends, and grades have to do with whether he graduates from home.

M: Yeah.

BK: Because he's too smart not to get a girlfriend if he really wants one.

D: T told him he was fishing in the wrong pond, and I kinda like that. (Laughs)

M: I do, too, but yeah, he's fishing in the high school pond instead of the college pond.

Act V: Bob Has Control Over His Life

BK: In Bob's world, he knows how to make things happen—including the illusion that he doesn't.

(S quietly laughs.)

BK: (Shaking S's hand) Do I know ya?

S: (Laughing) In a way ...

(M, D, and T all laugh.)

BK: (To Bob) It's all right. This just shows you're smart.

T: He is.

BK: Very smart.

T: Because his grades weren't good enough, somewhere in the 40s, 50s, and then I told the parents not to worry about him anymore so that he can use his own strength to work on it. And since then there has been an amazing performance.

BK: You know what's amazing? The amount he was able to improve his grades is the same amount of pounds that he needs to lose. Isn't that weird? You're talking about losing 40, 50 pounds, and he sort of zooms in and controls it just so that grades are 40 or 50. You probably didn't even know that you are that good. Bob's world can pull off anything.

In this act, we see that Bob can be contextualized not as having problems that are out of control, but as a young man who has exquisite, even uncanny control of his grades and presumably his weight and ability to have a girlfriend. He is not out of control but in control, and it has something to do with the family issues of his leaving home.

S: (Laughs) Stop flattering me.

BK: No, but you know this. This is too weird. How could that same number be there?

T: The last test, he was able to move from around 45 to 93.

F: You went from a 40 something, to a 60 something, and then a 90 something?

S: Yeah.

Act VI: "I'm Leaving, But You'll Never Lose Me"

BK: I'm going to make a suggestion.

I will now address how he can leave in a way that congruently and respectfully addresses the assumed contradictory family messages of "please don't leave" and "please leave and get on with your life." He needs to be able to say that leaving does not mean losing him.

S: Um-huh.

BK: This is something I think can be fully understood in Bob's world.

S: Yeah.

BK: That is, for every 5 pounds that you lose, you need to write a letter to your parents. It can be just one sentence saying "Don't worry, you'll never lose me."

S: Mmm.

BK: Because if you get the weight you want and you start getting the girls, they're going to worry about losing you. But you're always going to love them and you're going to come home and be appropriate on the phone and all those kind of things. This is how you've been stuck. From now on, for every 5 pounds you lose, just write them a letter saying, "I lost 5 pounds, but don't worry, you're not going to ever lose me." You know what I mean?

All the changes the family wants for Bob (weight loss, his getting a girl-friend, successful grades) come at the cost of Bob's leaving home. They want this for him but are struggling with his departure. This dilemma has been the underlying confusion. Bob has also mirrored the same contradiction by both acknowledging that there are advantages and disadvantages to staying and leaving. Now he is being advised that for every increment of change that he makes (and that his parents, especially mother, will notice), he is to assure them that he will always be their son and have a presence in their life.

S: You're a smart guy.
BK: You got it. (Standing and shaking S's hand.) Thanks, they're done. Okay. (Shaking hands with F and M.) Nice to meet ya'll.

When Bob said, "You're a smart guy" and nonverbally expressed that this was the family dilemma and resolution, I immediately declared the therapy finished. I exit to see what the family does with this very moment. How will they address departure?

F: Nice to meet you.
BK: You've got a real amazing son here.
M: Yeah, I think so. Thank you.
F: Thank you.
BK: A lot smarter than he even knows.
T: Thank you, Dr. Keeney.

(BK leaves the room.)

F: Does he teach here?
T: No.
F: Hum.
M: Is he visiting?
T: Every now and then he comes around to visit. Last year he came around and he did amazing things with …
M: That's cool.
S: He's an interesting guy.
T: He is, it's really a blessing …
S: He's very complimentary, even not knowing me.

(M laughs.)

F: No, he knew you.
M: He knew you.

T: So this has given us another dimension … Bob's going to work in his own world and whenever he is able to lose 5 pounds, he will write a letter saying "I'm not going to leave you."

S: Are ya'll actually worried about me leaving ya'll?

F: No, we know you're going to do it.

M: I've always been worried about it.

The parents present the two contradictory family positions about his departure.

S: No, I mean … I want to hear it … I know you have. I want to hear it from you (pointing to F).

F: What?

S: Do you actually care that I leave?

F: Yeah.

M: Well, because you've heard it so much that he couldn't wait until you left.

F: Now, I told you the other night when you come in there with your hands clenched and gritting your teeth mad …

S: I wasn't gritting my teeth.

F: I told you the other night I'll be glad when you're gone. But, no, I'll miss you. I don't want you to leave.

S: What is …?

F: But you've got to. You can't stay at home until you're 50 years old.

It's fascinating to hear the number "50" again. I wonder whether this particular statement of father's has been frequently articulated at home. If so, it would make sense that Bob's use of that number with his weight gain (and desired loss) and test loss (and desired gain) is also a metaphor for staying versus leaving home.

S: Was that out of anger or was it out of … ?

F: Yeah.

M: Yes … we wish you'd act nicer when you were at home.

S: You see …

F: And you're going to wish you did one day too.

M: Yes.

S: Maybe so.

F: You know, you will. But anyway …

S: Time out.

(Phone rings. BK calls T from behind the one-way mirror and requests that the family sharpen the focus on their son's leaving home. BK suggests that T say that Bob knows what to do, then proceed to stand up and attempt to terminate the session—all as a way of enacting the process of departure in the here-and-now of the session.)

S: Was that him?

T: Yeah.

F: Is he still in there (pointing to mirror)?

T: Um-hum.

S: Hi (waving to mirror).

(M and F laugh while the young D, who has been playing in the corner, gets up and also waves at the one-way mirror.)

T: In Bob's world, big things are happening. One is that he's writing a book [*Note:* He is writing a book and has told the therapist that he is a character in it.] and the other is that he's going to work on his weight. Lose 5 pounds and then write a letter. We know what to do now, don't we? All right. (Standing up) I think he's ready.

F: Think he's ready?

T: Yeah.

F: For what?

T: He's got what he needs.

F: You're turning him loose.

T: (To S) Lose 5 pounds and then write a letter to them saying that they shouldn't worry that they'll lose you. Isn't that wonderful?

(F and M laugh.)

T: It is.

S: Don't tell me you're leaving.

T: (To M and D) Are you happy for him? (M and D nod yes.) Yes, I am, too.

F: Where are you going?

T: You have solved your problems, so we don't need to talk anymore.

F: Oh, okay.

(M laughs.)

T: All right.

(Knock at door; BK enters.)

BK: Do you see what's happening? See how difficult it is to leave. This family is really struggling with leaving. This family is really struggling to leave.

We have staged an enactment of the family's struggle to handle the forthcoming departure. I will invite Bob to actually leave the family right there in the session, while promising that they won't lose him. This is what the situation is bringing forth—the successful performance of Bob's leaving home. Now we see the family also wrestling with leaving their therapist, mirroring the departure of Bob.

(F laughs.)

M: (Laughing) Yeah. We kind of took him (the therapist) into our family …

BK: Here is the reality: Your son is going to leave, but you're not going to lose him. He is going to leave. And you (looking at S) are going to leave now.

BK: And guess what? You're not going to lose us. You're not going to ever forget this session.

S: Two heads are better than one apparently.

BK: Yeah, because we've stepped into Bob's world.

(All laugh.)

BK: (Leaving room) Okay, have a nice life.

S: I will.

F and M: Thank you.

(D hugs T.)

T: Your sister likes it. She likes the fact that you are going to lose 5 pounds and then write a letter that they won't lose you. And she likes the fact that we are ready to go, right?

(M laughs.)

F: We're not coming back up here with you any more?

T: Oh, you can, you know, you can come and talk about how he is writing his letters and and that kind of thing.

(Phone rings from behind the mirror.)

Again the family is struggling with departure. I am calling to amplify the moment, to heighten its presence and importance in their life, to not let it dissipate away, to underscore that this is the moment for change and departure. I will ask the therapist to give a message to Bob.

S: He's back in there.

(M laughs.)

F: I don't understand what's going on.

D: (Waves at mirror) Hello.

F: I'm confused.

D: You're confused. (She walks right up to the mirror.)

Father, like Bob, is describing the confusion associated with this natural family transition, a time when contradictory messages about the desire to leave and not leave are being voiced and felt.

F: He's getting back in there pretty quick, ain't he?

S: He's moving fast.

    (D sits on F's lap.)

T: Ok, thank you. (T hangs up the phone.)

F: (Laughs) Go over there and sit down.

T: Maybe she likes to sit on your lap.

F: She does. She's a daddy's girl.

M: Oh, she's a daddy's girl.

S: She's a daddy's girl.

F: He was a daddy's boy until he got older. You know how boys are; they …

Note how father's comment about the son's growing up is relevant to the present scene of action.

T: (To S) Dr. Keeney has a message for you.

S: Uh-hum.

T: Since you know what you're going to do next, just stand up. Can you get up? (S stands up.) All right. Your parents are really worried about the fact that you will be leaving them.

S: Uh-hum.

T: Just give them the assurance that they are not going to lose you. And then you'll be okay to leave.

Act VII: "I'm Not Like My Older Brother: I Will Leave Home, But You Won't Lose Me"

S: I think I've told you this before: I'm not like Hank. Ya'll should know that. And you're about to cry … it's okay, go ahead. I'm not going to leave the family. I'm not what my older brother has become. I am loyal to this family. Yes, I don't visit Nana all of the time, but I have a life. I can't just all the time visit her. I got things to do.

We were very surprised to hear this pronouncement. We did not know that there was an older brother who had presumably left home and cut off all ties with the family. This was the first communication about it. Now we see why the family had such a difficult time preparing for Bob's departure. Leaving home had previously been a tragic event. Bob had been sensitive to this and had been creatively finding ways to address it through the metaphors of weight, grades, and success with dating. Now it is being voiced in a direct way that permits release and assurance.

F: I don't expect you to visit her all of the time.
S: I don't want to hear griping that I don't …
M: I don't gripe. I just ask you when was the last time that you saw her.
S: In a way, that's griping.
T: Okay.

(Phone rings from behind the one-way mirror.)

I want to help Bob get back on track, and so I ask to speak to him by phone. The therapist hands the phone to Bob and I tell him that he has done a beautiful job. I underscore that he has made it possible to leave the family in a way that his parents can now appreciate and accept. Now he must tell them that he will not disappear forever but that he must move on. After telling them that, I ask him to leave his family that very moment—to walk out of the session. On the one hand, this is a dress rehearsal for the real thing in life. It is also actually doing it right there and then. He says that he is ready to do it.

S: (To his parents) I love ya'll and I'm not going to leave ya.
F: Okay, we love you, too.

(S leaves the room. D gets up, closes the door, and moves into S's chair.)

M: But he left (laughs).
T: It's paradoxical.
F: Ah, he'll be back.
T: You are really worried about Bob leaving because you are shedding
    tears …

(Phone rings again.)

I advise the therapist to spotlight the moment. The phone calls, in themselves, also serve to spotlight, accentuate, and underscore the transformative moments. I advise the therapist to again simply remind them that Bob will leave, but they will not lose him.

T: Once he's turned 18, he's gonna leave you, but you're not going to lose
    him. This is what Dr. Keeney wants you to be aware of. Right?
    He will leave, but you will not lose him.
F: You hope.
T: I think we're done for the day, and I think your problem is solved
    because he's …

M: Yeah, I was thinking that, too. I was thinking, you know, this kind of helped me turn him loose.

T: Okay.

M: Not talking … about all that stuff … I've been on him since before 9th grade about his grades, you know, and if I had just known what I know now. I would have done this sooner.

T: All right, it is now over.

F: Thank you.

M: We're going to miss you.

D: I'm going to miss you.

T: You're going to miss me.

M: Yeah.

T: Oh, I will miss you, too.

M: We've told everybody about you. We just really, really like you.

T: Thank you.

F: Thank you.

T: All right.

M: We wish you well.

T: Thank you so much.

The family and therapist walked to the waiting room. It was an emotional scene of tears and further hugging as they all said goodbye. They also requested that I come out and see them one more time. They also wanted to say goodbye to me. I went out to the waiting room and joined in their celebration of getting on with the next stage of development in their family life. Bob went on to attend university, where he is a successful student who is enjoying becoming a young adult with a proud, supportive family. In a follow-up conversation with the therapist, Bob said, "I'm doing well and loving school."

CHAPTER 7
Cody

This session involves a middle-aged client (C) with a long-term treatment history for what was diagnosed as "schizophrenia," including residency in mental hospitals. The client, Rick, has also been clinically assessed as "paranoid" and sometimes believes that he is under the watch of the FBI. He is presently in therapy with Scott Shelby (S), a doctoral student at a university in Louisiana. This session was conducted and videotaped at a local social service agency. During the session, roofers were walking and hammering directly above the room, which made a tremendous amount of noise during the beginning of the session. We were able to get them to take a break so a quieter—though hopefully no less "constructive"—session could take place.

Act I: Therapy Street Smarts

(Session starts with a big crashing noise on the roof.)

S: We have roofers on the rooftop.

C: Oh, I was also looking around for the cameras but that is okay. Though I don't know this place very well, I'll trust the people here. Cameras make me a little nervous. ... I kind of lost my train of thought ... I'm all over the place today, really. Yeah, I feel like I'm talking about myself too much, but ...

S: Who would you want to talk about?

C: Oh, I don't know. You're trying to ask probing questions. I just want to talk about the library and how they are part of the conspiracy.

S: What about it?

(Knock at door. Brad Keeney [BK] enters.)

C: Hi, how are you doing?

BK: Hi, I'm Brad Keeney. Nice to meet you. Sorry I'm a little late.

C: That's all right.

BK: Fill me in a little bit. How many sessions have you guys had together?

C: Well, I was just talking about my paranoia a little bit and I guess we've had about one session.

BK: You've only had one session?

C: I think we've had two.

BK: Is this your first experience with therapy or have you had many experiences?

C: I've had a lot of experiences, but it's my first experience at this place.

BK: At this place? And with this therapist?

C: Yeah.

BK: And how often have you seen each other?

C: I think twice?

BK: Twice? Before that? How many therapists?

C: Oh, a lot. I'm sorry.

BK: That's all right.

C: I'm focusing on answering the question.

BK: That's okay.

C: A lot, a whole bunch of therapists.

BK: More than two?

C: Yeah.

BK: More than three?

C: Sure.

BK: More than four?

C: Over the past 25, 30 years, yeah a lot.

BK: More than five?

C: Okay, you've gotta stop.

I am setting forth the evidence that he has experienced many therapists and many sessions of clinical work. I will not use this to make the case that he is really sick and therefore required that much therapy. Instead, I will propose that he has become an expert at being in therapy. This different connotation of his past may lead us to an unexpected scene.

BK: Okay, so you're a professional.

C: Yeah, getting there. Trying to be, yeah.

BK: At this stage, you probably know more about therapy than most therapists.

C: Umm, I don't know. I try to …

BK: You can be honest with me

C: Yeah, but …

BK: But you've thought this a lot?

C: Yeah, I've had a lot of that thought. … I'm a little out of my depth, or off-kilter because I'm at a new therapy center, so …

BK: I understand that. I'm at a new center, too, so at the same time we'll both be on a different kind of wavelength with each other okay? You've heard of street smarts? I'd say you have therapy smarts?

C: Yeah.

Act II: On a Mission to Teach Therapists

BK: I'm having sort of a fantasy. Is it all right for me to share it? I've worked with a lot of therapists in clinics all around the world. I've also worked with a lot of healing traditions, too, in Africa, the Amazon, and so forth. I've learned to be sort of free with the thoughts that come upon me, but I ask permission if you want me to share the things that just pop into my mind. Something has just come into my mind: I'm wondering whether or not part of the mission of your life is to train therapists.

I am announcing that I will join him in allowing thoughts that pop up in my mind to be expressed in an unedited way. We will communicate in the same manner. With this established, I ask the wild question of whether his mission is to train therapists.

C: I'm not sure if it's my mission; no, it's not a mission.

BK: I'm not sure you know that this is not a mission. You know that when you get this smart in terms of understanding therapy, every therapist you meet is going to get educated or trained in a way they never expected because you're a client that knows a lot more than anybody else they're going to see in the whole history of their career.

C: Well, that's the whole, but I'm more interested in reforming the government right now.

BK: Yeah, well maybe you've got to learn how to reform the therapists before you have the skills to reform the government because they are not always that different.

I accept whatever he says and find a way to utilize it as a legitimate reason for his possible mission.

C: Yeah.

BK: So, you are in advance training for changing the world of therapy and if you get a successful mission on that front, you'll be ready to take on the government.

C: Well, I've tried not to obsess or overindulge at any of it.

BK: So what's your guide? Are you guided by principle or guided by something that's a mystical thing?

C: Principle.

BK: Do you have a name for that principle? Have you ever articulated that?

C: The principle is just, uhh, the conscience, my conscience, yeah.

BK: You feel when it's feeling right versus feeling phony?

One of the things I have observed about so-called "schizophrenic" ways of being is that lying is often an issue. This was on my mind when I made this comment.

C: Yeah, that's good.

BK: You were born with a highly developed—let's be nontechnical—and call it a well-tuned bullshit detector?

C: Yeah, I'd go with that.

BK: You just know it. You know what is real and what is not? That's a gift, but it's also a curse.

C: Yeah.

BK: It's like you walk into a place and all these alarms go off: bullshit, bullshit!

C: I also feel like I started this from the time of my birth. Well, not from birth, but from my young age. It could be I watched too much TV when I was a kid.

BK: It didn't seem to affect your natural gift of detecting bullshit. I have a theory that everybody is born with the same highly tuned instrument that can detect that which is not authentic, but somehow yours didn't get out of whack. Yours just sort of survived and now you're grown up and you still have the same sensors.

He has now been connoted as both a teacher of therapists and an adult whose bullshit detector still works.

C: Yeah, what's your name? I want to remember.

BK: My name is Brad.

C: Okay, Brad, cool.

BK: Okay, you bet. So, congratulations that's quite an achievement.

C: Well, thank you.

BK: I've met very few people in the world who still have their bullshit detector still intact, 100 percent accurate, and still working at your age. It's usually gone by age 5 or 6. You know what I'm talking about?

C: Uh-hum, yeah. I can feel what you're talking about, yeah.

BK: In other cultures, if you keep that detector alive, then you're respected as someone they go to in order to find out what's going on with others ... They trust them because they have good intuition.

C: Well, yeah.

BK: Maybe you're in the wrong culture? Ever think you're in the wrong country?

This is another way of saying that he has been in the wrong context—a setting that defines his conduct as crazy. Somewhere else in the world, his expression might be celebrated as opposed to pathologized. I am implying that he may need to move his contextual residency.

C: Well, no, but I'll tell you that I'm trying to deprogram from a lot of things. There was a communal experiment that I was involved in a long while back. It was pretty different; it was pretty wild.

BK: In Louisiana?

C: No, in California. It was very experimental.

BK: Hum? How old were you?

C: I was about 28.

BK: So you are even rarer a human being than I imagined ...

C: Well, rarer for this part of the country anyway.

BK: Not only is your bullshit detector fully intact, and operating well, you're a historical experiment.

Again, I am reframing any details of his life that may imply a problem or trauma. They are now rearticulated as unique and, in this case, "experimental." We now have a teacher of therapists, an adult with an operating bullshit detector, and someone who participated in an historic experiment. Note that for almost each time he is recontextualized, he asks me for my name. This time I respond in an atypical way to further advance the way we are joining one another.

C: Well ... and you're going to have to tell me your name again and I'm going to remember it this time.

BK: It's Brad.

C: Thank you, Brad.

BK: Just think "B"; call me "B."

C: Okay, so I'm going to call you Brad.

BK: Second letter of the alphabet. If you get to the second letter of the alphabet, then you'll remember that's the second letter following the first one, and after "R" is the third letter. B-r-a. B and A, one, then two: somehow you get that. I think weird things once in a while. I think it's good to think weird things and that's all right because I call that creativity. The trick is how to benefit from it. And make some money from it.

C: Yeah, that's a good way, too. … One of the things I've been talking to Scott about is that there's a lot of conditioning that society … more specifically, I'm talking about how rational recovery relates to addicts and helps get them over their addictions.

He reminds us of his therapeutic expertise, which I will subsequently acknowledge.

BK: Since you had so much therapy, you … should probably receive an honorary degree for how much you know about therapy, if that were possible, because you probably earned it through experience, and you probably read a lot of books about it, too. It wouldn't surprise me … But my concern for today is … if you had an experience that you never imagined, if you have an experience in this session that you will never forget.

Now that we are joined as colleagues—both teachers of therapists—I invite him to imagine having a more unique experience in the session.

C: Okay!
BK: Would that be something good?
C: Yeah!
BK: Would you like that?
C: Yeah.
BK: Or should we just bullshit?
C: Well, as long as it is not a Christian kind of dogmatic thing, you know?
BK: No.
C: Not all Christians are dogmatic.
BK: I don't like dogmatic of any kind. I don't like dogmatic atheists. I don't like dogmatic Christians. I don't like dogmatic therapists. I don't like dogmatic dogs. I don't like anything dogmatic because it's boring. You've got a mind that is sharp and you have an imagination. You know if it's not fresh, it's not authentic. That's your bullshit detector going off. Right?
C: Yeah.
BK: If you could have anything happen … lets just say that (loudly slaps his hands) something in this room just surprised us, startled us … a lightning bolt came into the room and all of a sudden we started talking in ways we've never talked like before, and it came out faster than we've ever imagined, and we started shouting at each other. Hey! Hey! Hey! We just turned things over and we said, "I don't want this room to ever be like another therapy room." (At this moment, BK jumps up and turns over the coffee table in the room.) I'm going to turn everything upside down,

even put the table legs this way. ... Let's both put our leg on an upside down leg. We at this moment have done something probably that's never happened in the history of psychotherapy. I bet this has never happened, so we are in new territory. Feels good, doesn't it?

C: Yeah, it feels good.

We are both enjoying the spontaneity of sharing our craziness in the moment. The difference is that I get paid for my craziness and he suffers from his eccentricity. I am showing him another choice, which includes owning that he is a teacher of therapists.

BK: Do you know I didn't know what I was doing? I just let myself be moved by creative inspiration that was inspired by your presence. I wouldn't have done this if you weren't in the room. I know that because you're sharp and that you detect bullshit, I've got to be real with you. I'm going to tell you right now, that I don't know a damn thing other than there's bullshit and there's also the real thing. I know that.

C: Well, you do know a lot.

BK: You can feel it. It's the feeling that comes forth that matters. Yes, I don't know. I'm just wondering whether you'll help me or whether I should help you—because you know so much.

C: We can help one another.

We are now colleagues and as such, we enter into a new act.

Act III: Cody

BK: That's good. I like that. I like that. You are real. That I feel. This guy is real. Yeah. Do you have a nickname? People ever give you a nickname?

C: No.

BK: What if someone from another planet came with higher intelligence and superior spiritual sensitivity, whatever that means, and gave you a special name. Do you know what happens in other traditions? You would be given another name for what you know. Without thinking, what new name for you comes to your mind?

C: Cody.

His new name is a metaphor that underscores his being a different person in a new contextual home.

BK: Cody, that's nice. Where did that come from? That's my curiosity ...

C: It feels okay, but ...

BK: Cody, I'm going to call you Cody. From now on, you're Cody. Would you name me? Give me a new name?

C: No!

BK: No? Amazing! Who is this guy? He's just so authentic. He knows that it just doesn't feel right to name someone else. You've got to come up with your own name, don't ya?

C: Yeah.

My own free association at the time was that he did not want me to abandon the way we were relating. His new name was associated with his being a colleague with me and there was no obvious need for me to leave the situation we were both enjoying together. A name change for me may have implied that exit.

BK: That's nice. Wow! I like Cody. It sounds Wild West.

C: Yeah.

BK: That's good. That inspires me to think of what you might consider doing in your life. Consider if you, here in this very moment, gave yourself, in serious fashion, permission to be Cody. You could construct experiences that are new for you. First, you might even get stationery that has the word on it. You could mail a letter from the past you to Cody. Everyday, you could read and write read letters from Cody. I used to advise clients to write a letter to God. Write a letter to God and mail it to yourself and see what that feels like. But I think your writing to Cody might be very interesting.

C: It might. I've thought about that before. … I'm very much trying to find out where my beliefs are at.

BK: Well, aren't you more interested in changing things before it's too late?

C: Yeah, I am, but I have to go with the flow.

Notice how Rick—now Cody—is correcting and challenging me as an equal participant in the dialogue.

BK: Okay, but I'm interested that you want to change the government and I had another free …

C: Well, Brad, let me stop you. I don't want to change the government; I want to change the Republican part of the government.

BK: I hear that. Amen! If we are going to be Christian, then let's say "Amen" to that!

C: Well, that's another thing. I'm exploring. I'm not a Christian.

BK: Well, this place is not a Christian organization. This is a social service agency that occasionally invites strange people like me to talk to people like you.

C: Well, that's okay. I have trust.

BK: Good. Whatever you want to change—whether it's the right wing element or people who run global and state and national government—one great way to get to them is through changing the therapists. They also go to therapists … You know that Republicans are all going to therapists; they have to be. How could you deceive and bullshit the nation and be in alliance with so many corrupt ideas and not be messed up and seeking a therapist?

C: I don't want to change the therapists.

BK: Hum …

C: I don't think everybody is messed up, but I think a lot of people are.

BK: So you're not as ambitious as I thought.

C: I'm not ambitious …

BK: Really? I had the opportunity to live with some of the folks who are the oldest culture on the earth, the Bushmen of the movie *The Gods Must Be Crazy,* where the Coke bottle falls from the sky. Do you know that movie?

C: I've heard of it, yeah.

BK: They are the oldest living culture. They've lived 30 to 60 thousand years and they never declared war on anyone. They are harmless people. They didn't have the kind of ambition some ruthless cultures have. … Do you see yourself as a teacher?

C: Intuitively.

BK: Do you write? Are you writing?

C: I try to write. As far as being antiambitious, I'm not antiambition. It's just that some people take their ambition too far.

After an episode of collegial talk about what he has been trying to change in the world, I ask him what he would like to change in his own life.

Act IV: Cody's Search for Peace

BK: I understand. What do you want to have different in your life tonight? One thing, small thing?

C: Peace.

BK: Peace. When was the last time you had peace?

C: Intuitively, four months ago.

BK: Four months ago? For how long?

C: Three minutes.

BK: Okay, so if you had four minutes of peace tonight, that would be an achievement in your life?

C: Yeah, I'll say 10 minutes.

BK: Ten minutes is okay. This is nice. I like this. This opens up a door to something we can work with. I think the first thing to do is to see if you can bring peace to Cody.

C: Okay.

BK: Forget about who you were. Let's think about what it would be like to bring peace for Cody. … If you were to help Cody achieve 10 minutes—well, maybe that would be too long—maybe it should be five minutes of peace. Could it include—please say yes or no—a good meal? Would that be part of it?

We will now build a concrete definition of what a small moment of peace would look like to Cody.

C: Yes.

BK: Would it include finding a movie that you would enjoy?

C: Yes.

BK: Could it include you writing one sentence that you are pleased with?

C: No.

BK: Would it include you doing something in the house that you have never done before?

C: No.

BK: So we are talking about a nice date. We're talking about a nice date and a nice meal?

C: Yeah.

BK: That's a date. Does anything not sound like a date you don't like? You need a nice date for Cody. That would bring peace. When was the last time you had a date?

C: Years and years ago.

BK: Maybe because it's taken a lot of your energy to sort through a lot of stuff. It doesn't leave a lot of time free … it takes so much concentration and concern and there is a heavy burden associated with your former responsibility …

C: Yeah, you have to be able to let go.

We have a turning point that suggests how he can move forward—letting go of his worries about taking on responsibility for many big issues. I spell it out for him.

BK: My sense is that you have taken on responsibilities that have fallen on you because of your talents, your bullshit detector, the skills you have in psychotherapy, and that which you believe needs to be done out of a good conscience.

C: My thing is to worry too much about things.

BK: Yes. You need a recess. You just need to build in a recess. If you don't get a recess, you are not going to be successful doing the things you feel are important. The prescription that I'm going to write for you is for you have to have a recess.

"Peace," a word often applied to world conflict, has been traded in for "recess," something more accessible to an individual.

C: Okay.

BK: A recess means that you take time to not do any work ... And for you, you have already defined what a nice recess would be. It's a date. Why don't you have a date with Cody? Have a date with Cody. Have a fine meal and watch a fine movie.

C: Okay.

BK: That would be a recess. Can you do that? Are you creative enough to set up a situation where a ...

C: I am, but it might take time.

BK: Aw, that sounds like you're sort of on the edge of uncertainty. Why not tonight or tomorrow night?

C: No, because I have some things to do and I'm too involved with my family.

BK: When can you find time to find a recess? This is what it's going to take.

C: Saturday.

BK: Saturday. I would suggest that you consider what would be the appropriate food or meal for that day or evening. Will you go out or stay home?

C: Can I write this down?

BK: Absolutely—you should write this down. I need to go get some water. Why don't you guys work on this? Write out in detail what would be a great date for Cody in terms of what food and what drink? I don't know if you're a juice guy or a water guy? Whatever. Plan the whole meal. Will there be a candle? What color candle? Where will it be set? Will it be a picnic? Will it be a restaurant? Will it be at home? Will it be in your room?

(BK leaves the room.)

Rick is now looking at a way to have recess—making a date with Cody. All of this "date" talk also points the way to being more involved with the world. I left the room for all of this to ferment inside Rick's mind.

C: Well, my intuition is kind of going wild right now.

S: I bet. Well, what have you written out so far?

C: Just that it's Saturday and ... it will work out.

S: What would you like to have for dinner, or would it be lunch?

C: I don't know.

S: Would you like to do it at dinner time?

C: Yeah, I would.

(BK reenters the room.)

BK: So did you get your date planned?

C: We're getting there.

BK: Are you going to do this?

C: I'm going to try to do it.

BK: What would it take for you to say to me right now that you are going to do it?

C: That's interesting because one of the fellows at the Solomon house used to say: "Trying is not it. You don't try. You *do*. You don't try."

BK: You see you're teaching right now. Now that's something Scott can use.

S: That's remarkable.

BK: Yeah, I'm having a fantasy that when you have this dinner, you might— this is going to sound really weird—but you might imagine having an empty chair around the table for every therapist you've ever had.

C: That's an idea!

BK: After all, you've been a big part of their lives. You've probably been training every one of them.

C: Well, the thing is, Brad, is that a lot of them, including the psychiatrist that I've seen, were not very good people.

BK: Yes, I can understand that. But you also met some others who were ready to learn from you.

C: Yeah.

BK: But, I want to back up, because it's easy for me to get distracted because you are such an interesting guy and can take me off in other directions. I really want to hold the line in the sand. What would it take to get you, without being overly ambitious but sufficiently motivated, to actually pull this off? I think if you have a date in the way that we are talking, a date for Cody ...

C: Yeah, I was hearing a date as in "a date with a woman."

He, in his imagination, is moving out of psychotic isolation into a world of imagined new relationships.

BK: Yeah, well, the Cody I was talking about: Cody is the new you! You know what I'm saying. I'm calling you Cody. As of this very moment, in this very session, when things were turned upside down, you became Cody. I believe that. I've lived in cultures where people get a new name. Inspired by the moment of receiving a new name, they walk away feeling they are reborn. They feel like they've become somebody else. At least they've made that commitment.

The dramatic moment when I turned the coffee table upside down and we placed our feet on top of the bottom of the table legs is marked as the moment his world changed. He then started to enter a new contextual home, upside down from his previous residence. Here, as Cody, he is competent and capable of training therapists, engaging in creative dialogue, eating out, seeing movies, and thinking about dating girls.

C: I wish I could do that.

BK: Well, it takes a little drumming, a little singing, a little dancing. Do you know what I mean?

C: Yeah.

BK: We could do that if you like. I'm a pretty crazy character. I'd bring a drum in right now and drum a song for you and give you an initiation for you if you'd like. Would you like that?

C: Well …

BK: Is that too much for you?

C: That's too much for me.

BK: I figured. You're not that crazy!

He's now less crazy than my presence!

C: My mind is kind of wondering.

BK: Is your mind sort of struck by a possibility that emphasizes that if you really wanted to … if you really wanted to sort of baptize yourself, I don't mean in the Christian way, I mean the renewal [of your life].

C: Yeah.

BK: If you really want to be made different, then you really must decide to be in life in a fuller way that includes a recess and has some peace.

C: I have a thought.

BK: Yep.

C: The thing is that I'm thinking of atheism and the rituals that go along with that.

BK: Yes. Atheistic rituals?

C: Yes. And I'm just not there yet. I'm going off in too many directions.

BK: That's all right. Be patient with yourself. Right now, I'm only worried about you having a new name, the possibility of 10 minutes of peace, and a date with Cody. That's Mount Everest right now. I want to see you get to that summit before you take on lower mountains cause you have to start high before you can do things lower. That's why you turned things upside down. You know what

I'm saying? You're the kind of person that has to start at the top of Mount Everest before you can achieve successful ascensions to the small mountains where everybody else has their problems.

C: I feel like I'm trying to control a bit too much.

BK: Saturday night: time for a date with Cody.

C: All right.

BK: Appetizer, main course, dessert, choice of beverage, with or without candle, and the movie.

C: Well, I'm going to work at it.

At this time in the session, as we move toward his commitment to new action outside the therapy room, I challenge Rick to leave the room as Cody.

BK: What are the odds? Greater than 50 percent? Less than 50 percent?

C: Probably is less than 50 percent.

BK: 40 percent?

C: Yeah, I'd go with that.

BK: How do we increase it to 60 percent? What do we have to say to increase it to 60 percent?

C: I have to get some sleep first.

BK: It's too much to have it at dinner? Maybe you should have breakfast?

C: Yeah, that'd probably be easier.

BK: Let's change it. Make it breakfast then. Go with breakfast. Instead of a movie, maybe just a small cartoon for sure.

C: Okay.

BK: A cartoon and breakfast. Maybe that's too much. How about a glass of juice and a commercial?

C: Well, I'll give it a try.

BK: Okay. Would you agree with me that you will accomplish one of these three things? You'll either have dinner and a movie, or you'll have breakfast, or you'll have just a glass of juice and a commercial.

C: Okay. (There is a long silence in the room.)

At this moment, I thought that Rick was imagining how his world could change. As someone who typically talks nonstop, he was now silent, lost in his imagination, as if seeing his future.

BK: Have you ever sat in a room with a therapist and no one ever said a word? There was just silence.

C: Oh, it's okay.

I will now ask Cody, a teacher of therapy, to help me. We will be equal colleagues present for one another.

BK: Funny how that silence gets loud. I'm happy you're on the planet. I get discouraged myself these days.

C: Well, I don't try to get discouraged and I'm totally discouraged, but I'm not going to let that get to me.

BK: If I turn on the news, I get depressed.

C: But there's a lot of people getting involved …

BK: Can you tell me something that will help me with my attitude?

C: Yeah.

BK: Your expertise is obvious because you've been so serious about these matters.

C: Okay.

BK: Do you know that I have not been bullshitting you?

C: I know that you're doing what you need to do, yeah.

BK: It's all about having a bullshit detector …

C: Brad, I'm a little concerned about the cameras, but that's not something I'm going to control right now at all.

BK: Well, fuck the cameras. Right? Right now it's you and me. If anything, the people who watch will simply learn from you. It's your gift to them. I see you here in a different way. I see you going to all these sessions as a teacher. I'm just going to bring it up again to remind anyone who's watching. That's why the cameras are here.

Even his paranoia is brought inside the contextual home where Cody teaches others. In other words, in this setting if they are watching, let them learn something!

C: The university was a good experience and it was a good place to teach.

BK: Good. Will you write us a letter and tell me what happens? 'Cause I'm a guy who needs some hope.

C: Okay.

BK: If I see you able to do something with your life, it's going to give me hope.

C: What can I do to help give you more hope?

BK: Just know that what I want to hear is that Cody is alive and well.

C: Cody is alive and well.

BK: And that Cody is not only having peace, Cody's having some fun. Furthermore, Cody is making some good kind of trouble—stirring things up in a good kind of way. You know what I mean?

C: Yeah. One way or the other, I have that built into me and I have learned to just survive.

BK: Excellent. I think that you have an interesting life ahead of you, including the new ways you will consider training others, and

> liberate and find those little spaces from time to time that give you peace and release those.

C: Okay, I really appreciate very much your involvement.

BK: You bet, Cody. Thanks man. Do they hug people here?

C: Yeah, they sure do.

BK: That's good. I think we're all wrapped up.

(C and BK hug.)

Four months later, I received an e-mail letter from Scott Shelby, the therapist in this session. He wrote:

> Brad,
>
> I wanted to update you on the case we worked with while you were here. It is the case of Rick/Cody—I hope you recall the experience. Here is the follow-up from July 16, 2007. It is very exciting!
>
> Rick called and told me that he is no longer living at home with his mother and brother. He said that he has decided to live out of his car until the final arrangements can be made on an apartment or house trailer. He announced that he will live by himself, though he is nervous about being on his own for the first time in such a long time. ... He gave me the assignment to study William Glasser's therapeutic model. He also discussed some limitations he sees in the approach of AA and how it doesn't always fit certain personalities. He has remained in the frame of a teacher and I have continued to be his student (in a therapeutic way). ... Rick said he was on his way to the mall to catch a movie. He hadn't decided which movie he would watch, but he needed to get off the phone so he could get there on time.
>
> I am in the process of writing my dissertation on the evolution of your clinical work. I would like your blessing to use this interview to describe how one conducts therapy in a creative, improvisational, resource-focused way. I think this session is a great illustration of this approach.
>
> Thanks for your time. We all look forward to working with you in the fall.
>
> Scott Shelby

Nine months later, I received another e-mail letter from Scott about Rick. Rick had stopped by the clinic to say goodbye. He was on his way to San Diego, California. He had been living in a trailer and was maintaining an income and a checking account, which he stated was "a big deal for me; I've never had one of those." Rick talked about all the progress he had made and how comfortable he was with himself. Rick's last words to Scott included these remarks: "I appreciated my conversation with Brad last year and would like to meet with him one last time. ... Brad was one of few people who ever connected with me."

CHAPTER 8

Peashooter Love

This session involves a husband (H) who raises pigs and hires himself out for general carpentry and his office-working wife (W). They came for therapy because of constant quarreling. The co-therapy team consists of male and female student therapists (TM and TF, respectively). As I observe from behind the mirror, the husband talks about how he has been helping his neighbor build a barn and how he's tired when he comes home. He mentions that he works all day and doesn't have anyone to talk to until he comes home, but his wife gets home too tired to talk. She needs her "down time" after work, but her husband counters with, "She acts like she's the only one that works. You come out there and handle that lumber all day. ... She used to come down there and work with me." Looking directly at his wife, he goes on, "That lumber's hard, ain't it?"

His wife shakes her head and, with a sigh, speaks about their dilemma, saying when she gets home from work and sits down to close her eyes and relax, her husband walks in and says, "Open your eyes!" She then gets up and lies down in bed. The couple proceeds to enact the beginning of a disagreement.

Act I: Wired-In Automatic Arguing

H: A lot of times, I'm just playing with her, picking on her, but ...

W: And I tell him those are not the times to do it ...

H: (Sighs) When is a good time to pick? (Laughs)

W: I don't know.

H: We used to just wrestle around and go down on the bed and I would just tickle her. Now she says she's not ticklish. I enjoyed it then.

That's what we used to do. But now I'm scared to hell of doing anything. I ain't gonna lie. She just miffs at me and pops off.

W: He always says I get mad, but I don't always get mad. He takes it that I'm mad. (Her hand gestures indicate, "What can I do?")

H: Now what does our oldest son say to you 9 times out of 10 when you talk to him? "Why are you so mad; why are you so grouchy?"

W: That's what they think.

H: "Why is Mom so grouchy?" That's what they ask me, and I say, "Don't get me to lie. I don't know why she's that way."

W: Maybe it's my tone of voice. I don't know. But I'm not mad and I'm not upset but …

H: Just like coming over here to therapy today … I asked her a question … and then she rolled her eyes. (H moves his head and rolls his eyes.)

W: I did not.

H: I could see the eyes roll back and I said, "Whoops, time out, excuse me."

TM: Um-hm.

H: I get that. (Points to W, who is rolling her eyes.) See? (Laughs) See? I guess that's really why we're here to see y'all 'cause we got up here and we was going this away (hands are together) and then we went that away (fingers point in different directions).

W: So what if I get upset? I'll get over it!

H: Yeah, hell, it takes two weeks to get over it!

H: There we go again. (Sighs) I try too hard to be where she won't be mad (emphasizing his words by regularly pointing his fist toward his wife) … I'm scared she gonna get mad. Well, she will get mad at the (he snaps fingers) drop of a hat. I notice it. The kids notice it. I asked her about it coming over here. I got that look. And, I immediately said, "Oh Lord, mistake number 100."

W: That's part of our problem. … He worries that I'm going to be mad or I'm gonna be upset. … I'll get over it! And then he'll say, "Well, I'm gonna change, I'm gonna change." I'm not asking for change.

H: I ain't gonna lie to you. I AM UTTERLY SCARED to say something to her. It didn't used to be that a way. … I was in the military and someone could say anything, but I went on with no problem. But with her, I'm scared. I'm scared it's gonna be mad city all the way.

The couple has shown me the habituated way they are locked into quarreling. It is time to intervene. In other words, their performance is asking, even begging, for calibration. In this session, I will use interpretation and explanation as part of the performance and as a means of setting up movement toward an alternative staging of their life.

(There is a knock on the door. BK enters the room.)

TM: Hey, y'all, this is Brad.
W: Now we're in trouble.

(BK shakes H's hand.)

H: Hey, he ain't got on a Hawaiian shirt! (Laughter) How you doing, sir?
BK: (Shaking W's hand) Hi, I'm Brad.
H: They said you had on a Hawaiian shirt.
BK: That's interesting, isn't it?
H: Yeah.

(BK sits down.)

BK: Let me ask you both a question.
H: Oh, oh …
BK: Do you think you're the first couple to have this situation?
H: Um-um.
W: No.
H: And won't be the last.
BK: So you know … (H nods) … that this is part of the marriage show.

I am defining their situation as a routinized drama that many marriages get stuck in. Doing so both normalizes them and defuses any pathologizing. This is a step toward loosening the grip their habit has on them—giving it less authority than has been presumed.

H: Um-hm.
BK: Okay. That's important to know. That's actually very good. (I look at TM and TF.) That's a very important thing to know.
TM and TF: Um-hm.
BK: Because a lot of married couples think they're the only ones who get stuck like this. Or they think they're stuck in a way that's very different than anyone else.
TF: Um-hm.
BK: All right? And it's like quicksand. (H nods and starts to open his mouth to say something, but BK continues.) But both of you want to be out of it. …
H: The more you move, the more you sink.
BK: You got it. (BK looking at W) And you feel set up. (BK now looking back at H) I mean you're angry before you've done anything to make her angry. (H nods.) Because it's so automatic. (BK looking at W) You can't help it. You know what I'm saying? I mean,

just when he starts, it's like you think, "Okay, you're setting me up, and now I'm already angry though I'm really not; I don't want to be angry, but I'm already getting set up." That's how it feels on her side. And you (looking at H) know how it feels on your side because you've been talking about it. You both know.

I spell out how their quarreling is an automatic repetitive sequence—that is, a vicious circle. As the husband defined it, "The more you move, the more you sink." He has proposed that we can utilize a pig farmer's version of the "interactional view" formerly articulated in psychotherapy by Wátzlawick, Weakland, and Fisch (1974).

H: Yeah. It's getting to a time where I see 3:30 coming around each day…
BK: Right.
H: Ah, 3:30 is when she gets home. 'Till then, I have a good day.
BK: She's probably got the same pattern. (BK looking at W) You probably start worrying exactly at the same time and you're already seeing in your mind how it's going to happen. It's almost like you're dress rehearsing. (Everyone laughs.) As soon as you see each other, it no longer matters whether you've promised that, "I'm going to do something different." (BK now looks at H) No matter what she says about, "I'm not going to say things I usually say." (BK turning to W) No matter what he says about, "I'm not going to get upset." The very second you look at each other—you don't even have to do anything—it happens.
H: You're right; you're right!
BK: That's it, right?
H: Yeah. You're right about that.

Now that we have defined the dance they are stuck in, it's time to move forward and find an alternative choreography.

Act II: Messing with It

BK: You want to know how to get out of it?
H: Close your eyes? (Laughs) I was just messing with you.
BK: That's good!
H: How old are you?
BK: How old do you think I am?
H: Well, I'd say 52.
BK: I wish. I'm 57.
H: Oh God, I'm older than you.
BK: I bet I've seen a lot more folks in this mess than you.

H: Oh, in this. (Everyone laughs and BK shakes H's hand.) Yeah! But
ah ...

BK: But, just a sec, what you just said was the secret.

H: What's that?

BK: You said "messing with you."

H: Oh yeah.

BK: You all need to "mess with this." The problem is you're not messing
with it. You're not experimenting with it. You're not tinkering
with it. You're locked into it. You're expecting it. You can't stop
it. Believe me, you can't stop it. But, you know what? You can
mess with it.

Utilizing the pig farmer's continued explication of an "interactional
view of psychotherapy," without his knowing that he is embodying any
therapeutic position, we will define the situation as requiring some "messing with it."

H: Yeah.

BK: If you mess with it in the right way, you can short-circuit it. This is a
whole different thing that is not ever thought of because you're
too locked into the way in which it appears so rational to both of
you from the side from which you see it.

H: Well, you know, ... we used to go up to Diamond's Bluff ... you go right
around to the end and look straight down and you see a lake—
it's way down. I mean you can see faraway up there. We used to
go up there and sit and look. I've asked her to go up there 'cause
it's like the old habits.

The husband is drawing a distinction between their "old habits," which
were resourceful, and the present ones they are stuck in. I will not pursue
this example of what may seem to others as a possible solution or problem
exception but stay with him in the present moment, reminding him of how
they are now on autopilot for arguing at 3:30. Why? We are in a contextual
theme that is bringing forth movement toward "messing" with their present dance. This movement will carry us somewhere, and we won't stop to
indulge in a fantasy about the way things used to be. Note that his consideration of a positive relationship came forth as we shifted to a situation
where we are able to consider messing with the present dance.

BK: Yeah, but [now] you [would] still have a fight ...

H: Yeah.

BK: At 3:30, you still worry about it. It'd be a nice thing to do. It would be
really cool.

H: Yeah.

BK: I'm talking about the times when it's happening and how you might tinker with it. You ever have a peashooter? (Both nod.) If you used a peashooter when you got angry, you'd be messing with it.

In the very beginning of the session, the husband described his contribution to their dilemma as follows: "A lot of times, I'm just playing with her, picking on her." His wife responded that, "those are not the times to pick," to which he countered, "When is a good time to pick?" What metaphor involves both "playing" and "picking" and a "right time to pick?" In these hills, it could mean some good old-fashioned peashooting!

TM: What's a peashooter?
BK: Tell him what a peashooter is. (Laughter.)
H: You don't know what a peashooter is?
TM: I have no clue what it is!
H: Next time you go to McDonalds, get a straw. Then use a little BB. (Puts his hand to his mouth and blows through it.)
TM: Ah ... gotcha.
H: Boy, how old are you?
TM: (Laughs) Huh? I just never did that!
H: Boy you led a sheltered life.
TM: I need to bring one in here.
BK: It's old school. (BK turns to W) But you know what I'm saying?
W: I get you.
BK: I mean you could throw corn seeds or you could use a peashooter. I don't know what it's going to take to short-circuit this thing. But you know what? I know that neither one of you can stop it by trying to stop it.
W: Um-hm.
BK: But your only chance—and I'm talking to you as someone who has written over 30 books, has directed clinical programs all over the place, has given many lectures, and seen thousands of people—I can tell you this right off the bat: you're not going to be able to stop it by trying to stop it. But you have the choice to mess with it, to try to short-circuit it. I don't know what that's going to be. It's going to require your using a little imagination and it can't be logical. Anything logical is just going to make it worse. So maybe trying a pea ...

Further validation and contextual solidification of the importance of being inside the frame that prescribes "messing with it."

H: Illogical.

BK: … a peashooter. Try to aim it at his knee. It could be as simple as that.
(BK looking at W) Or the second you get angry, go bake a pie,
but bake it with something that he won't like.

H: (Looking at W) We know about them pies, don't we?

BK: What's that about the pie?

H: I can make them pies! I make the pies.

BK: Maybe you ought to make the pie.

W: Nope.

H: She don't like me interfering in there.

Act III: Vacation Time

BK: Right now, I would like to invite you … (H slaps W's leg and then BK
looks at H to focus attention.) I'd like to invite you …

H: To go to Hawaii?

BK: I just came from Hawaii.

H: You did?

BK: I was in Honolulu a couple of weeks ago.

H: Oooo …

BK: It's nice there.

H: (Nods) I'll tell you what would cure my …

BK: So you know what the two of you need?

H: A vacation?

BK: You need an Hawaiian state of mind. An Hawaiian state of mind says,
"I need to take a vacation from trying to stop this thing." From
this moment on, I don't want to hear any more talk about your
dilemma (BK looking at TM and TF) because I know it won't be
useful. Don't even try to say anything rational about this arguing thing.

We have moved to the next act, defined as giving them a vacation from
their interactional habit, including not trying to stop it or think about it
with any rational means.

TF: Mm.

BK: 'Cause it's just locked in like it is for many married couples.

TF: Same thing.

BK: (Looking at the therapists) They try to stop it because they're both
good people (TM and TF nod.) They both want to stop this.
They're both wanting to make this thing work, but it just auto-
matically happens because it's wired.

Their previously unsuccessful efforts to stop their problem, when seen
from inside this interactionally described frame, have paradoxically

maintained it. Now the same interaction that was connoted as problematic is resourcefully designated as an indication of good people who want to make things work out for their marriage.

TF: Um-hm.

Act IV: Short-Circuit Your Habit

BK: It's got to be short-circuited. Short-circuiting means adding something that doesn't fit. (BK now looks at both H and W.) When it starts, or just before it starts, or while it's going on, or at anytime, throw something into it that doesn't fit. It might be a peashooter.

H: (H looks at W.) I know how to use them, too.

BK: It could be a peashooter.

W: (W looks at H.) So do I.

BK: Really?

W: And I've hid the batteries to the cattle prod (holds out her fist as if indicating she is holding a battery).

H: Oh. (H shakes his head and laughs.) You ought to know better …

W: The cattle prod.

H: I got a cattle prod, um …

W: That would not be part of it.

H: I had some …

BK: A cattle prod! (Everyone laughs.)

How fascinating that when we moved inside the theme of short-circuiting their habit, the electrified cattle prod is brought out! Be careful about thinking and talking about "circuits!" Yes, we will utilize the cattle prod.

H: I had some hogs out there. They weighed twelve, fourteen hundred pounds apiece. I had to use it on a couple of them. After I sold 'em, I brought the cattle prod up to the house. Well, once when the oldest boy and the youngest one were chasing each other around the house and making havoc, I spoke up and says, "The first one of y'all who bumps into me might see that cattle prod." I shouldn't have said that.

BK: (Looking at TM and TF) You know my fantasy?

TF: What?

BK: Give them both a peashooter. But his has to be a straw attached to his cattle prod. (Laughter. BK then looks at H.) When you use the peashooter, you gotta hold the cattle prod and have that …

A peashooter attached to an electric cattle prodder is an exceptional metaphor for a means of short-circuiting a farm couple's stuck situation.

H: I, I, I …

BK: No, no, no. See, the fact that you're hesitating and wondering whether you can do it tells me this might work. If you thought that this would work, I'd know it couldn't work. Now you can't use the cattle prod; you're only going to have the peashooter attached to it and then use the peashooter.

His stuttered words are immediately seized as an example of hesitation that indicates uncertainty, the very thing needed to affirm that it qualifies as an action that might short-circuit their being stuck in a looped instant replay. In other words, we are pointing to something that is outside the box of how they have understood their interaction, a classic old-fashioned second-order change.

H: I know …

BK: (Addressing the therapists) If they thought this made sense and was a good thing, I wouldn't think it had a chance of working. They must first think, "Oh, I can't imagine doing that." That tells that we might have something to experiment with it. Maybe she ought to have a frying pan with her straw attached to it.

W: I have a cast-iron skillet.

Each spouse is now fully armed with a peashooter held by an exaggerated instrument of aggression, stage props that assure an absurd encounter of the most ridiculous and short-circuiting kind.

H: There you go.

BK: You must have some rules about this.

W: You can't do it when I'm asleep.

H: I'll go to sleep. (W laughs.)

BK: Are you gonna sleep when she comes with a pea …

W: I gotta a cast-iron skillet.

H: I'm gonna be awake …

BK: Let's say you did this. Let's assume you get your cattle prod and put a straw on it and that you get an iron skillet and put your straw on it. Now the next time it's 3:30, you're going to say to yourself, "I know it's going to happen, but this time I'm gonna get my peashooter." How we gonna do this? You'll need to get in two opposite corners of the same room. Then see if you can actually shoot the other all the way over.

I'm setting up the stage to maximize absurdity.

H: Yeah.

BK: That'd be one thing. I would say that if you get hit that far away, it should be something that's not going to hurt, but you'll know that you were hit by it. Maybe it shouldn't be a pea ...

H: (Whispering to W) A BB gun might also work. (W laughs.)

They are now teasing one another. Their habitual interaction is already being short-circuited.

BK: This is what's going to happen ... (blowing through a curled up hand, using it as a peashooter ... when somebody gets hit. When you are hit, just yell as loud as you can about how much it hurts you. Say in an exaggerated way that it hurts you so deeply that it makes your feelings hurt. ... I'm going to tell you that if you can do this, it is going to get short-circuited.

I'm suggesting an exaggerated parody of their present habitual responses to one another. A Leaning Tower of Pisa is being built, getting taller and taller, until it topples once the belly laughter commences shaking it down.

W: (Looking at H) Put Boo Boo [their dog] out first.

BK: What can you put in there besides a BB?

H: Ah the usual. You know, chewed up paper; put a little bit a piece of paper in there ...

BK: While standing in two opposite corners, it's going to take a little skill to hit the other person.

H: You get a grain of rice. Can't do that, though.

They already are agreeing to soften what they throw at each other, all metaphorical indications and enactments of change.

BK: (Addressing the therapists) Now you all discuss it, but if they start to say, "Yeah, that sounds easy," then be distrustful.

Again, I'm trying to keep them away from a first-order framing that is indicated by "that which makes sense to them."

TF: Mm.

BK: Right now I think they're in that zone where it's like maybe it might, and ah ... Let me stop and ask you, do you really want to stop this?

H: Yeah, I wouldn't be here. (W nods in agreement.)

Act V: Pointing to Their Love

BK: Okay. You're willing to mess with it, which means tinker, experiment, and short-circuit. (Both H and W are nodding; BK stands up, shakes each of their hands and then puts their hands together.)

Promise? (The couple laughs together and continues to hold hands. BK points at their hands.) Look; look at that right there. Look at that hope. That's called hope. Right there. See, that's called "love" right there. You see what you just did?

I am spotlighting a dramatic moment by standing and pointing to their hand embrace, raising my voice and proclaiming that this is something important. This heightened moment of the session takes us toward the love they both hold for one another. I will continue with this crescendo of noticing and enthusiastically framing their affection.

H: Oh, I know, I knew she …
BK: You two love each other.
H: I'll tell you what …
BK: See you two love … look at that! Look at that right there! Just behold it …
H: I …
BK: … without saying a word …
H: I know …
BK: … BEHOLD THIS (looking toward TM and TF). You look at this right here.
TF: Um-hm.
BK: That's all you need to know. Help them short-circuit. I'll be back. (BK starts to leave the room.)
H: Hey doc? I'll tell you one thing. She can tell you this. There ain't been a person, even her daddy, that says I don't love her. I tell 'em I love her and she's mine. (There is a knock on the door and a delivery of two straws from the director of the clinical graduate program, who had been observing from behind the one-way mirror. Everyone laughs together.)

The couple, now expressing love, is now fully armed with peashooters delivered by the highest ranking official in the clinic.

BK: Awesome. That's for you.
H: (Looking at TM) Hey, now this is a universal peashooter!
TM: Peashooter! I got you now.
BK: You know what? The man who delivered those straws is older than both of us. He's the director of this whole place.
H: Oooo!
BK: These peashooters come from the head honcho. Now you all talk a little more about how you're going to set this up and I'll be back in a bit. Okay?

(BK leaves the room.)

H: I like him.

TM: He's good.

TF: You got your peashooters there.

H: Yeah.

TM: You got a cattle prod and a frying pan ...

W: The batteries are hid.

(They proceed to talk about setting things up until BK reenters the room.)

TF: Come in.

BK: I just can't get over what happened there.

H: What's that?

BK: How you just reached out for each other, held your hands, and showed me your love.

I am continuing to validate the dramatic moment when they touched and held on to one another.

H: (H whispers something to W that is inaudible to everyone else, and she immediately hits him on the chest, as if he has said something naughty.)

BK: Hm? (The couple laughs.) Look how easily they laugh. They got the love; they got the laughter. I know you are going to do this. We shook on it. ... As soon as one of you feels the trouble starting to come, even before you open your mouths, get your peashooters. Get your cattle prod and frying pan, and you gotta promise me that you'll each go to a corner of the room. Because if you're too close, you know ... I want you to be distant. I want you to be distant from each other 'cause that's what ...

I am underscoring one resource after another from affectionate expression to laughter, all naturally present in the contextual home they have moved into. Here the peashooting is framed inside affection and laughter, love, and absurdity. It has a different meaning when performed inside this act, a stage that holds and brings forth a connected couple. I will rearticulate the forthcoming dynamics of the peashooting dance so it is voiced and experienced on this new stage.

H: You gonna take all the fun out of it.

BK: No, you must get distant because the fun is when you actually get the other one. When that happens, the other one has to whine like a baby. Okay? The one who is hit by a paper wad has to really whine like a baby. ... Do you have a camera?

Again, I am exaggerating their responses to being hit by a paper-wad pea to make it absurd. Now we will tie things up by a utilization and extension of the dramatic moment that took place when they held hands.

H: Huh? Yeah.

BK: I'm gonna ask you one more thing. This is in addition to what we already talked about. I'd like for you to hold each other like you did earlier and have someone take a picture of that. (H playfully aims the straw at W's face and tries to blow the wrapping off.)

Peashooting is already taking place, not as a way of fighting but as teasing and joking with affection.

H: It didn't work. I missed.

BK: Okay?

H: Cheap peashooter.

BK: I'd like for you to take a picture of you each holding hands like you did right now.

H: What?

BK: Just that. The photograph must show the two of you holding hands. Are you following me?

W: Yeah.

BK: Okay?

H: Real good.

BK: Do you know what I want you to do with that picture?

H: Put it in *Ripley's Believe It or Not.*

BK: I want you to put that photo under the middle of your bed without thinking why. Because every night when you go to sleep you're going to be sleeping over the two of you holding hands. And your mind is going to wonder, "What does it mean that we're caught in this kind of dance with each other while we really are two people that like to hold each other's hands." Okay?

They will sleep over their remembered affection, a way of taking this scene with them. A photograph will frame the moment when they left the stage for arguing and stepped upon the stage of love, and place it underneath their dreams.

H: All right.

BK: Have somebody take a picture and put it underneath your bed. Think about it when you go to sleep. And, boy, there's going to be some peashooting fun. (H again tries to blow the wrapping of his straw toward W.) All right. All right. Nice to meet you. (BK shakes hands with H and W.)

H: Nice to meet you, doc.

BK: It really is impressive the way you can move (BK rhythmically snaps his fingers with an ongoing beat) from the giggling to being ready to square off with those peashooters.

I am implying that, with a steady beat, like a smooth dance, everything can happen naturally and spontaneously—they can get as entrained to the rhythmic movement of a virtuous circle as a vicious one.

H: Tinker with it.
BK: Tinker; yes, now you're gonna tinker. All that other stuff's past. Now you're gonna tinker. Great. Take care.
H: See you.
BK: You bet.
TF: Thank you.

(BK leaves the room.)

H: I like him.

(W smiles.)

TM: I do, too.
TF: Um-hm.

(The director [D] of the program enters the room again, but this time with a Polaroid camera.)

H: Oops.

(BK comes back into the room.)

BK: Guess what we got here? You got the chief. (Everyone is laughing.)
D: Can I get a picture of you with your straws and this loving process you're in? Is that okay? (H puts straw in his mouth and W grabs it.)

We are freezing the scene: capturing the image of their being in the last act of this therapeutic performance and sending it home with them.

BK: Oh yeah. Excellent.
D: We caught you …
H: See she grabbed mine …
D: We caught you loving, is that okay? (Wife nods yes.)
BK: That's good.
D: All right, here we go.
BK: Why don't you hold hands? Hold hands. (H and W hold hands.) Perfect.
D: There you go. Okay. (D takes a photograph of them cuddled close, holding hands.)

BK: (BK moves close to W.) I bet the way he caught you originally was that
     he's a big teaser. Was he always a teaser?
W: Yeah.

I now am able to frame the very things that were previously problematic, his "picking" and "playing," as "teasing," the very thing that caught
her attention when they romanced one another.

BK: That's how he courted you.
W: I don't know; I didn't pay attention too much. (The couple laughs.)
BK: Yeah. (D hands the developed Polaroid photograph to W. BK then gets
     close to H.) Your wife's response right there is how she caught
     you. She pretended she wasn't paying attention to you.

I utilize her response of pretending she doesn't notice or pay attention to
him, which was what irritated him in their arguments, as being that which
originally attracted him. Now that they are in a context of affection, Act V,
they are attracted and pulled together by that which repelled them in the
beginning scene. As always, the art of awakening a session is flipping contextual homes, getting the same behavior to move residence to a context
that is resourceful and transformative.

H: You are telling the truth. That's right.
BK: That's how it works. All right … This is what I call "interesting
     chemistry."

Their chemistry is interesting because in one setting it has them verbally
shooting at each other in a painful way, while in another theatre, just the
thought of peashooting helps bring on affection, laughter, and closeness.

(BK and D start to leave the room.)

H: Hey, your camera ain't gonna work anymore.
W: It worked.
TM: It worked? All right. Good. We all ready to tinker?

(Everyone stands and shakes hands.)

The couple has been short-circuited by their affection, love, and
laughter. The latter came forth when there was a stage erected for its
performance. Creative therapy is as much about being a producer who
finds an appropriate stage as it is about the subsequent action. When
the stage is set, props are at hand, and the lights are on, all one has to do
is open the curtain and the performed lines will come forth effortlessly
and naturally. In a creative theatre of transformation, any surprise can
enter, including peashooters and a paparazzo with a camera in hand to
catch the actors in the act of expressing love. What does this couple's

future hold? No one and no outcome study can really fully know the future of any client or whether therapy had anything to do with whatever happens. But we do know that this couple went home with two peashooters, a short-circuited habit, and a photograph of the moment when they changed their performance to a different stage. They left with some peashooting love, a resource they can always draw upon and shoot away any circuit that needs some prodding.

For the record, the couple returned for one more session and said that their life was now "better" and they felt no more need for therapy. They said they had been "shooting" each other and the wife was shooting the most, saying she enjoys it and found it effective in "stopping" her husband's "assumptions" about her. They commented on how much they enjoyed the peashooting session and they were happy with the way everything turned out.

CHAPTER 9

Seaman's Song for His Son

This is a session with a mother (M), father (F), and daughter (D) discussing their 18-year-old son and brother, Bob, who is presently incarcerated in prison. He'll be getting a temporary release from prison for a home visit. Dad works as the captain of his own boat, which brings cargo to oil platforms at sea. He says when he is out, all he can do is fish, clean the boat, and play cards because there is no television reception when he is so far away from shore. He can't even swim because of sharks. Mom says he's been doing this for 30 years. "In the old days, it was wild; now you have to be concerned about safety." He took his son out to sea with him before he began serving time. The session takes place outside of New Orleans.

Act I: Is it Bob or Robert?

Brad Keeney (BK): I understand your boy's coming home?

M: Um-hm.

F and M: I, we hope so (laugh).

BK: Is that a certainty?

M: We believe …

F: He's supposed to go in front of the judge either the last of this month or the first of next month.

BK: He's doing all right?

M: Oh yes.

BK: All signs are positive? Have you thought about how you're going to welcome him home in a special way? Have you given some consideration to this because there has been a big change.

M: I know it.

BK: He's become used to another place and …

M: Right.

BK: … and he's more than likely a different person.

M: Um-hm.

BK: Because he's grown and had some experiences.

F: I'm hoping he is.

BK: Even by his age he's not the same …

M: Yeah.

BK: … He'll be a different person.

F: We can tell he's changed.

M: Oh yeah.

F: Just by going in there and talking to him and stuff.

BK: Really?

M: Yep.

BK: Good. So you're going to have a new son.

M: Yep.

BK: How do you welcome a new son to a home? Do you give him a new name?

By listing all the ways their son has changed—getting older, living in a different place, new experiences, observed changes—I now explore how their relationship with their son will change, doing so by addressing what name they will call him when he visits home.

M: (Laughs) Yeah, Bob.

F: No.

BK: What do you call him? What's his name?

M: I call him Bob.

F: We all call him Bob.

BK: What's his middle name?

M: Robert's his middle name …

BK: What if you start calling him Robert?

M: Hm, I don't know.

BK: Instead of Bob.

(Father laughs.)

BK: Seriously.

M: Um-hm.

F: (Talking over BK) I'll call him Bob. Hell, ain't nothing changed.

While previously acknowledging that Bob has changed, Dad is asserting that his relationship with him won't change. I will push him further on this matter.

BK: If you call him by the same name he used to have, he might think he's
the old person.

M: I still call him Robert …

Now Mom says she also calls him by another name. Keep in mind that
the family lost a "boy" to prison, who will soon visit home as a "young
man." They haven't gone through the transition of experiencing this change
with him. It is no surprise that they are confused as to whether they are
relating to a boy or a man.

F: (Talking over mother and raising his voice) I'm just going to call him
Bob. I ain't changing what I call the boy because he got in
trouble.

M: No …

F: (Raising his voice to a shout) I'll call him BOB! I'll call him just what I
always called him.

M: Bob.

F: And every night, I'll call him Boo.

Dad is holding onto the name he used to address his son when his son
was a child. He still wants to hold onto that relational posture.

M: Yeah.

F: 'Cause that's what I've always called him.

M: When he was little. Boo.

BK: Very interesting. But you know what I'm saying …

M: I will. I'll call him Robert …

F: I'm gonna call him Bob.

BK: Yeah?

M: If you gonna call him Bob …

F: (Very angry at this point) There ain't nothing changed in my eyes. He's
the same boy who I raised.

The parents are enacting their differences and confusion over how to
relate to their son. It is feeling very intense in the room. Observers watching the session from behind a one-way mirror later acknowledged that they
were feeling stress just observing the father's escalating anger. I am seeing
his emotion as a possible resource that is naturally being brought forth, so
I am allowing it to amplify.

M: Yeah.

BK: Right.

F: (Now with the strongest possible emotion in a loud, wavering voice)
Now if you got us in here to think I'm going to change the way I
treat the son I love (F stands up and moves toward me as if he's

ready to punch me) I'll walk my ass right on out of here right now.

The attending therapist later, after the session, said that he thought I was going to be knocked out at this particular moment.

BK: (Relaxed) No. I think it's good.
M: (At same time as BK) No, it's just …

(BK offers his hand to shake F's hand. Surprised by the calm invitation to shake, F shakes BK's hand and starts to shift his emotion away from anger.)

F: It's like you're saying …
BK: (Talking over father) I think what you're saying is good.
F: … you want us to change … I mean we didn't do anything wrong. (Sitting down)

Parents sometimes feel vulnerable to being blamed for their children's behavior, knowing that some people see kids in trouble as a consequence of faulty upbringing; however, in this session, this is less important than the parents' confusion over how they will relate to their son. Rather than explore whether deficient parenting was an issue, I will underscore their voiced concern and spirited expression for one another.

Act II: A Father Who Loves His Son

BK: (To therapist) You know this shows a father who really loves his son.

I am connoting father's emotional display as a message of love for his son, which moves us into another act. The drama served as a fulcrum to teeter us toward another contextual place.

F: I love my son. I'm gonna call my son what I've called him since the day he was born.
BK: That's good. And he is your son.
F: All right!
M: Yeah.
BK: He'll always be your son.
F: (Talking over BK) I'm not gonna change what I call him just 'cause he got into trouble.
M: No.
BK: No, we're not saying that.
M: 'Cause he knows it's Robert.

Mom is saying their son has changed and grown into a young man, and that father knows it.

F: 'Cause he's paid his price and I'm hoping and praying he learned something. So, I'm going to call him the same thing I've always called
   him.

BK: (To M) But you know what I'm saying.

M: Yeah. Sometimes I do call him Robert, too.

F: (Talking over mother) I know what you're saying, but I'm not going to
   change what I call him.

BK: No, you shouldn't.

M: 'Cause I'll call him Bob Robert …

This sort of says it all: they are stuck between relating to the son they
knew as a boy and the one they now don't know (how to relate with) as a
man. This is true for all families at this stage of development, but more so
for parents who have lost contact with an incarcerated son.

BK: When a boy is small, they usually say "daddy."

M: Yeah. (F nods.)

BK: Then there comes a time when they say …

M: Father?

BK: (Nodding) Dad or father.

M: Yeah, yeah.

BK: Okay. But what does he call you?

M: Daddy.

F: Dad.

Again, their relationship is confused—they are jointly talking about a
man-child named Bob Robert.

BK: Okay. You say "daddy" and he says "dad."

M: Daddy. He calls me mom, mama …

F: Dad, pops.

F and M: Pops.

M: Yeah.

BK: But you know what I'm saying. There comes a time when …

F: Oh, I realize. Like I'm saying, I'm not going to change. I mean I raised
   the boy right. He made some mistakes. But the way I raised him
   is not why he's in there.

M: No, he's not at all involved.

F: I'm not going to change what I did, what I do …

BK: Yeah.

F: … just because he's now away from home a little while.

BK: Yeah, but you …

F: If he joins the army and he goes overseas for four years and he comes
   back, I'm gonna treat him the same way.

BK: You know what I think?

F: (At the same time as BK) I'm not going to change how I treat him.

BK: I think if he saw how you …

F: I mean I love the boy.

BK: (To M) I think if he saw that, he'd be touched, don't you? (To F) What you just did …

F: Well, I mean, I'm sorry if I offended you or anything …

Act III: A Father Willing to Take a Stand for His Son

BK: No, you didn't offend me. You know what you did? You impressed me.

F: I'm not going to change what I do just 'cause he got in trouble.

BK: You know what you just did? You said I'm willing to take a stand …

Father's dramatic stand when he shouted at me is reframed as a loving father willing to stand up for his son.

M: Yeah.

BK: … for your son.

F: Well, I always have. … And like I told him, "You done wrong but I mean I'm still going to be there by your side."

M: Yep.

BK: Does he fully know that?

F and M: Oh yeah, he knows that.

BK: Great!

F: And there's something else he knows. I'll never bail him out of jail.

Now he shifts to addressing how he treats his son as a man rather than a child.

M: Um-um.

BK: (Looking at therapist) Maybe this has something to do with why their son has done so well in prison.

F: I think …

BK: Because he knows that he's got a father who will stand up for him … his father is going to stay all the way with him … he knows that the family's there for him.

M: Yep. The whole family is.

BK: That might have a lot to do with it. But, at the same time, how long's he been there?

F: Over a year. Right?

M: Yep.

BK: Did you say that he seemed to change?

M: Oh, yes.

F: Oh, yes.

BK: Okay.

Act IV: A Loving Family that Needs a Welcome Home Party

F: You can tell. I think what's got a lot to do with it was that his uncle, my
   brother, passed away.

M: Yep.

F: 'Cause we all lived in the same house.

M: And he was like a daddy to him.

F: (Overlapping conversation) I mean he's been there since Robert was
   born.

M: He didn't have kids.

F: He always spoiled him and that one, too (pointing at D).

BK: Really?

F: He just died and they wouldn't let our son go to the funeral.

M: Um-um.

F: He was actually like another father 'cause, I mean, he was there from
   day one.

BK: So there's been a lot of care, a lot of love and support.

M: Oh yes.

BK: Let me start again. I failed to see how much love and caring there was
   in this family. I'm sort of knocked out by how strong your stand
   is for your son. I'm going to ask the same question, but now I'm
   going to ask it even more seriously. How are you going to welcome this boy home? He's been away for more than a year? He's
   over a year older, and he knows that (looking at F) some things
   are not going to change. You're always going to be there for him.
   Yet, he's been out of the world and that's not an easy transition.

We return to the invitation to celebrate a new beginning (of now relating
to a man rather than a boy), but discussing it within the frame of a family
that expresses great love, is willing to take a stand, and does not have to
assume that they are being questioned about having failed as parents.

M: No.

BK: He's soon going to graduate.

M: Yep.

F: Yeah. He got his GED.

BK: Has he had a graduation party?

Another metaphor for the relational changes implied by a transition
from a boy to a man.

M: No. But that'd be something to think about when he comes home.

F: I hadn't …

M: His Ma Ma and Pa Pa [grandparents] would come.

BK: Did he miss a birthday party since he went to prison?

M: Yeah.

BK: (Looking at therapist) So he's missed a birthday party …

M: That's two things right there. Yep.

BK: He missed a graduation party.

F: Christmas.

M: Christmas.

BK: That's right. And did I hear right that he's going to start a new job …

M: On the boat.

BK: … with you? So that's a good thing?

F: Yeah, he's supposed to go on the boat.

BK: You know, sometimes people have a party for getting their first job.

All these reasons for celebration are markings of change in the family's development cycle, a boy ready to leave home and begin life as a young man in the world.

M: Yeah.

F: We'll probably have a kind of a welcome home party.

M: Oh yeah.

F: Like her mama might come up.

M: Yes, Ma Ma and his nanny.

F: We'll end up probably going out to the grave.

M: Yeah.

F: First day or two.

Act V: Father and Son: Composer and Poet

M: He writes letters for us to go take to his uncle.

F: (Overlapping conversation) He writes letters for me to take out there and read all the time.

M: He wrote a poem while he was in there and (pointing at D) she read it at the funeral.

BK: He's a poet?

M: It was sweet.

BK: He writes poems?

D: He writes good poems.

F: He impressed everybody at the funeral.

BK: Really?

BK: Where did he get that from?

I'm looking for a way to positively connote the family as being the source of Bob's creativity.

M: My sister. She writes poems and …

F: (Overlapping conversation) I don't think he got it from me.

BK: You're not a poet?

F: Not really. I wrote a few songs, but I …

BK: Really? Hold it, hold it. You guys are … it's overwhelming me now. You mean to tell me there's a poet in the family that inspired him and there's a songwriter?

Having moved past any suspicion and concern the family may have had about their deficiencies, we are contextualizing them as loving and creative.

M: Yep.

F: (Laughing) I ain't got nothing published.

BK: What kind of music do you play?

F: Radio.

BK: Yeah, but you write songs …

M: Lynyrd Skynyrd, Hank Williams …

BK: Really? You play guitar and sing, too?

F: Kind of country, southern kind of …

M: You see, right before Bob got locked up, he bought him an electric guitar. He wanted to learn how to play guitar, too.

BK: Really?

M: Yeah, 'cause he went out on the boat with his daddy for 14 days and made a thousand dollars.

BK: Wow.

M: Cash money. I've saved his check.

BK: (To F) How many songs have you written?

F: Four or five.

M: Five. Yeah.

BK: This is something that just comes out of my mind, because I just sort of speak what I think and sometimes I don't think, it's like …

M: Yeah.

BK: … all kinds of crazy stuff comes out. Is that okay? If I just say things?

(Everyone agrees.)

BK: I see the good that's going on here. You ever thought about writing him a song?

F: Yeah, I've tried. I've wrote something and sent it to him. It was some lyrics, but no music.

M: Yep.

F: It wasn't quite a song.

M: No.

BK: That really is touching. If I came home from prison and my dad wrote me a song, that would be the greatest gift (F laughs and nods) I ever received.

M: Something you'd never forget.

BK: What would be even more amazing, but maybe it's too much to even ask, is if the whole family sang it.

(Family laughs.)

M: Yeah.

F: Yeah.

BK: Wouldn't that be something?

M: Yeah.

BK: If all of you sang the song he wrote.

M: Uh-hum. Yep.

BK: You ever do anything like that?

F: No.

BK: Would you, could you do that?

F: I don't know.

BK: … that's the sort of thing I'm thinking about.

M: Right.

BK: 'Cause this moment …

M: Yeah, 'cause he's grown now …

BK: He's made his mistakes.

M: He's not no baby.

Mother is able to say he is a man without father correcting her.

BK: You told him the truth. You told him "This is what you need to do, and if you don't, you're going to get into trouble." He got in trouble, and then he learned.

M: Yep.

BK: He's now there. He's grown up. He got his degree.

M: Yep.

BK: Now there's this important moment that's like the hinge of his life. It could go one way or another—you don't know …

Becoming a responsible man versus an irresponsible adolescent is the fork in the road the family faces right now.

M: That's right.

BK: … but the moment he comes out, he'll sort of be suspended between having been in prison with those regrets … and the future.

M: Um-mm.

BK: He's got a future. The present moment is the most open he may ever be in all the time you are ever going to be with him.

I'm marking this moment in the family's history as a rare moment, an opportunity for growth, change, and transformation.

F: Right.

M: Yep.

BK: If he comes home and dad, or daddy, has written him a song from his heart and you all sing it and record it …

M: That's right. (F laughs.)

BK: I think that may be the anchor for him. No matter where he is on the sea of life, no matter what storms are blowing him around …

M: Yeah.

BK: … he'll have that song. It'll be his solid ground and he will always know that his family will be with him in that song.

We have moved from a child who needs a father to stand for him to an adult who was given an anchor by his family to enable him to stand on his own ground.

M: Yeah.

F: I'm gonna get to work on that.

M: Yeah.

BK: Would you do that?

M: That would be nice. Yep.

BK: I mean really. Could you commit to that? I'm asking you because …

F: I probably could …

BK: You'll never get this …

F: … I mean I got plenty of time to do it.

M: Yeah.

BK: 'Cause you know why? This opportunity never will come again.

Now is the time for the family to make a resourceful developmental transformation, finding a way in which they can actually be brought closer together rather than split apart when their son becomes a man. The opportunity is now, when he is Bob Robert, suspended between being an adolescent and a man.

M: That's right.

BK: As he gets older, he won't be as open. He's now between being an adolescent and becoming a man. Moving from getting his degree, going home, going out into the world. … This is the time.

M: Yep.

BK: It'll be a gift. It'll be a medicine. It'll be a power if you wish, you know, because the power of love in a song is the greatest thing.

M: Yep.

BK: You know how that works.

M: Um-hm.

BK: Okay. That would be awesome.

M: It would, yeah.

F: (Overlapping conversation) That's nice. ... That's a good idea.

M: Um-hm.

BK: Can you get them to sing it if you write it? Would they do that?

M: Yeah.

D: I probably could.

M: You could?

F: Yes, she probably would.

BK: Good!

F: Because she's the one we had to get up there to sing at the funeral and read the letter.

The family is organizing to accomplish this production that celebrates Bob's growth, while bringing them closer together.

M: You would not believe this poem he wrote for his uncle.

BK: Really?

M: It's just awesome.

F: Yeah.

BK: Wow.

M: We got it. We saved it.

BK: (Overlapping conversation) You know what that says? Your son is sensitive.

M: Yes. He is.

BK: He's sensitive ... and that means that for him, more than other young men you can imagine, he would be touched *more* deeply by this kind of gift.

(Family nods.)

BK: You know what I mean? Other boys will be touched, but he will be *deeply* ...

F: Next time we meet, we'll bring that poem to you.

M: Yep. Sure will.

BK: Excellent.

F: 'Cause I mean everybody at that funeral ...

M: Oh yes.

F: ... they all wanted copies of it.

M: Yep.

F: They were saying, "I can't believe ..."

M: That it came from him ... out of his heart, you know.

BK: (Looking at the therapist) I think the boy's okay and he's strong because they've done a good job. I don't think we need to worry about him. I think the most important thing now is not to solve any problems because there's really not a problem to solve.

Bob's becoming a man is a way of validating the family rather than breaking it apart.

M: Nope.

BK: There's nothing that should be changed. … Right now is one of the most important opportunities for this family's life. They can focus their attention on Daddy writing the song and singing it, and maybe even recording it, so when he comes out, this is the gift he'll have. Whew!

In this act, "nothing should be changed" means that the family should remain loving, supportive, and creatively expressive.

M: Oh yeah, I'm going to have the song, the poem, the cake, everything.

BK: Give your boy that gift from your heart …

M: 'Cause I miss him. I love my boy, too. I miss him.

BK: Mm, mm, mm, mm, mm. (M is crying and D puts her arm around her mother.) Actually, that really moves me. All right. Look, both of you, mom and dad showed me how much love there is in this family. Mm, mm, mm. That kid's got everything a kid could ask for, you know?

F: Yeah.

BK: (To therapist) I think they love him so much that they ought to give him something extra. And that's a song. That's it. I don't need to spend any more time with you. You've impressed me by how much you care. You've got the gifts and you've got something special you could do for him. Do you know what? When you do this, it's not only going to be a gift for him, it's going to be a gift for you all.

Whatever they do that is best for the son will also be what is best for the whole family.

M: Yep.

F: Yeah.

BK: When you're together and sing that song, you're going to feel it, and it will bring you together in a way you'll never forget. Okay? Are we all right with each other?

M: Yep.

F: We're cool.

BK: All right (stands up and shakes hands with F). I'm very, very impressed
with you, man. (F laughs.) You get it. I love my son, too. I know
what that's about.

I identify with the seaman as a father, anchoring us to the same
ground.

M: Thank you.

F: I think we just had a little misunderstanding.

BK: It's all right.

F: I didn't know what you was trying to do …

BK: You know what that earlier episode did? It showed me who you really
are. I would have never known. We couldn't have gone where we
went unless I knew who you were.

What could have been a disaster in the beginning (some observing therapists said they would have left the session and called for help) was utilized
as a way of accessing the family's strength and concern for one another.

F: Well it was just like when we went to court. I wasn't going to bail him
out and I didn't. I told her (pointing at M) that he will go to jail.
I'm not going to bail him out.

BK: You drew the line.

He drew the line distinguishing Bob, who is now becoming an adult.

M: Yep, that's right.

F: And the judge, he said, "What? You ain't going to bail him out?" I said,
"No sir!" He said, "Well, look, I'm going to release him to you.
You take him to work with you."

Perhaps the judge knowingly or unknowingly said, "If he's a man, then
he needs to work as one."

M: Yep.

F: That's when he went to work that two weeks.

M: Yep.

BK: Now this is a crazy thought, but I can imagine the song having a lot of
lyrics and one lyric is, "I went to therapy (everyone laughs), that
man said something, and I stood up for my son."

(F smiles and starts to head toward the door.)

Everything in the session is utilized as contributing to the family's new
production of a family song for Bob.

F: All right, we appreciate it.

M: Okay.

D: Thank you. Nice meeting y'all.
BK: Take care, you all.
F: Yeah. We'll see y'all next time. All right?
BK: Nice to meet you.

Bob Robert had a huge homecoming party. Friends and family gathered to feast and hear newly composed music and a few recited poems. Months later, he was released from prison and went out to sea with his father. There the two of them, with their two guitars, are writing poems, lyrics, and music about life at sea and visits home. Inspired by his tough father's tender song, the son has become a sensitive, but resilient, singing poet at sea.

CHAPTER 10

Spiritual Eyes for Trance

This session is a consultation with a therapist (T), Shannon, who is a faculty member of a graduate counseling program and his client (C), a young woman recently married who is struggling with a career decision. She believes that there is a "right path" for her life and that she needs to be focused on finding it. As she puts it, "I accept that there are many paths," but "my gut feels there is one that is right for me." She has a deep spiritual faith in her Christian tradition and the therapist, a middle-aged man with associations with Native American spirituality and Christianity, has been addressing her with spiritual metaphors. She and her husband, who is not part of her therapy, are outdoors people who hunt, fish, and camp out quite frequently. They are getting ready for a trip to Africa.

Knowing that Shannon's previous sessions had been using hypnotic ways of storytelling, I was encouraged to use a natural trance induction. In this session, the client and the therapists go into an altered state of consciousness as her situation is discussed. I immediately start working with the primary metaphors I heard her mention while I had been behind the one-way mirror. I advance each utterance in response to her body cues and agreements. The conversation is between how I loosen and open her ideas for more possibilities and alternative choices as she affirms with her smiles, head nodding, tone, and positive utterances, whether the direction of talk is meaningful. Shannon and his client have been chatting for a few minutes and he has already introduced me (BK) as we pick up on the action.

Act I: Many Paths, One Right Road

BK: Nice to meet you.

C: Well, thanks; this is great.

BK: You're very lucky—he thinks just like I do. You're going to get a double-barrel session.

C: Fantastic.

BK: Isn't that something? I'm really intrigued with what on the surface appears to be a choice between many paths, or many roads, and then considering whether there is only one.

I am creating a possible opening by differentiating surface knowing (many paths) from another way of knowing where diverse paths become one rather than separate.

C: Um-hum.

Act II: Spiritual Eyeglasses and Spiritual Vision

BK: What my mind or my inner voice tells me is that this isn't an either/or. Sometimes it appears like there are many lines, but that's because it's out of focus. When you get things focused, all those many lines can become one line. Once I had a dream that I was in an airplane flying very high and it had a hole, rather than a door, in its side. I was in a line of folks who had to jump out as if we were parachuting, but we didn't have any parachutes. In the dream, a voice said, "You're going to be reborn now." I thought, "That's interesting. I'm going to jump out of a hole in a plane flying high in the sky." When I looked out the hole, I saw many lines. The voice went on to say: "You have to bring all those lines into alignment until you see them as one." Then I sort of squinted and said a little prayer, like, "Help me have enough spiritual sight. Show me the way." When I said that and focused, those lines converged and I knew I was to jump. I just followed that one line all the way to the ground.

Here I use an actual dream I experienced years ago as a metaphor for her situation, emphasizing that many paths and lines can converge when seen through a certain focus or level of seeing. Seeing from an airplane high in the sky is higher (spiritual) seeing, where you jump with faith rather than a parachute and are able to use spiritual seeing to guide you through the complexity of differences that hover over the world below.

C: Hum.

BK: When I heard you were talking in a similar way, it made think of that dream and it made me want to come in here and share it with you. It's not necessarily a choice of one line or the other. It may be a choice of how can you squint, concentrate, or put on spiritual eyeglasses, or turn on a switch for spiritual sight—because that's a different way of seeing so everything converges. Interesting, isn't it?

The client's either/or distinction between many lines versus one line is reframed by saying that any line, when seen in relationship to all the other lines, may constitute being the one special line. Seeing this line requires a different kind of seeing, squinting, concentrating that requires spiritual eyeglasses. I have shifted the discussion to spiritual vision.

C: Um-hum.
BK: Interesting that you would talk that way and I would have this experience to share.
C: Um-hum.
BK: And that we would be here at this time and be able to share this connection.

I'm implying synchronicity or a significant coincidence in our meeting to further validate our being in a spiritual frame of reference.

C: Um-hum.
BK: It's like you bringing in a line and he's talking about a line, and I come in with a line.
C: Um-hum.
BK: Yet, we're all three together.
C: Um-hum.

I use the experience of the moment with three people in the room who feel together in their focus and spiritual framing, though different as individual people.

BK: It's interesting, isn't it?
C: Um-hum.
BK: Already something is happening in the room.

By addressing the spiritual frame or theme we have entered, its intensity is raised. The room felt more like a solemn spiritual occasion rather than a clinical exchange. I utilized this to bring us even more deeply into the new contextual home.

C: Um-hum.
BK: Because I can feel it.
T: Yah.
BK: We each brought a line in.
T: Yah.
BK: Now somehow we're here for a good reason—to teach and celebrate the mystery of life.

C: Um-hum.

We have rooted ourselves to being inside a spiritual framing (teaching and celebrating mystery) for considering her life and our interaction.

BK: Sometimes things appear to be in opposition, but it's just that we're out of focus.

Her way of being stuck between whether all paths are valid versus there being one correct path is redefined as an out-of-focus view.

C: Um-hum.
T: Yah.

Act III: Anointment

BK: What is it that you really want besides your getting your calling anointed?

I am proposing the metaphor of "anointment"—that is, a spiritual sanctification or a unique spiritual framing—as another way to discuss her situation. This is another step further into her world of spiritual understanding.

C: Um-hum!
BK: Or is that it?
C: That's all.
BK: You want an anointment …
C: Um-hum.
BK: … of your calling.
C: Yah.
BK: Good, because that means you're 90 percent of the way there. Most people don't realize that's what it is they most deeply want.

By agreeing that she wants a validation of her spiritual calling, I am able to imply that this very desire is the path she is looking for and that she is practically there.

T: Yah.
BK: Some people think that there's some textbook answer or they think it's all about talking with a career counselor. … What you want is an anointing on your life.
C: Yah.
BK: Your *heart* just simply has to say, "I want it."

I have associated "head knowing" (textbook answers and career counseling) as distinct from "heart knowing" which comes forth when you ask sincerely for another way of being.

C: Um-hum.

BK: And say that, "I am present, able, ready, willing." Then your part's done.

The heart line she desires comes forth by asking for it, something she presumably has already done many times in her devotions and prayers.

C: Um-hum, um-hum.

BK: Your part is done. There is nothing more than that. You simply ask and it's anointed.

Again, she asked for the blessed path, and that is all that is required for it to be present.

C: Um-hum.

BK: How are you going to know that you're anointed? You just have to turn on the switch that says "put on spiritual eyeglasses now."

I change her dilemma from looking for the right path to seeing that she is already on it. It is matter of tuning into the spiritual vision that confirms her sanctification.

C: Um-hum, um-hum.

T: Hum.

Act IV: Get a Spiritual Eyeglass Case

BK: (To T) I recommend that she get an eyeglass case.

T: Yah.

C: Um.

BK: Imagine that within them are some spiritual eyeglasses. They're not going to be visible to you or others.

C: Um-hum.

BK: But will be something you will feel.

The world of spirituality is not seen, but felt. This implies that although her mind can see many different paths and lines, her heart can feel there is one.

C: Um-hum.

BK: When you feel like there are many choices and many roads and you feel that all of these differences make you a little confused, know that you're stuck in your head. Remind yourself: "I need to put on my spiritual eyeglasses."

I am further redefining her situation: when she gets spiritual feelings that indicate that things are out of whack, make a visionary adjustment—put on some spiritual eyeglasses. Now we are talking about both spiritual feeling and seeing. What is not spelled out is that spiritual seeing is feeling.

C: Um-hum.
BK: You'll feel the truth of what you know is most true.
C: Um-hum.
BK: Then put them on. Look into the world saying, "Yes, I see different choices, different roads, but because I have my spiritual eyeglasses on, I feel one with the world."

Here her spirituality is fully utilized as a means of creating resolution when she feels stuck or confused about making what appears to be a myriad of choices.

C: Um-hum, um-hum.
BK: I know it's one.
C: Um-hum.
T: Yah, I feel it.
BK: Me, too, I feel it. Can you do that?
C: Um-hum.

Again, the feeling in the room is acknowledged as verification, even sanctification, of the spiritual truth of her life.

BK: That would be awesome.
C: Um-hum.
BK: Maybe you'll become a new kind of person who tells others about sanctified seeing.
C: I think so.

Telling others about sanctified seeing is a metaphor for walking a spiritual path.

BK: Spiritual eyeglasses. You could start with that. Perhaps you could actually market them in some way or maybe you could give them away. I don't know. I'm sure that there are many people that have come to this point of wanting their life to be sanctified. They say: "I want my purpose, my calling, my road, my highway. I want that to be one with the mission of my life: congruent, fully embodied within the Holy embrace."
C: Um-hum.

BK: I want to be on the Lord's highway.

C: Um-hum.

BK: In the little cotton field churches, they say, "I want to be on the highway to heaven. I want to go up that highway to heaven." That road. It is going to make you excited when you feel it. Because you feel it.

The session has become a church service. Her therapy is being blessed or sanctified by her most important beliefs and feelings.

T: Yah.

C: Um-hum.

BK: Having those eyeglasses—I'm going to tell you something—a long time ago among some spiritual people, including the Quakers and Shakers, there were those who felt that they received spiritual gifts like eyeglasses.

T: Um-hum.

BK: They did. But it's a feeling thing. Perhaps it's part of your calling to tell others about this. You know there's spiritual sight.

C: Oh, yah.

BK: But we need to go through a ritual to remind ourselves that it's there.

It was never about her finding new knowledge but about reminding herself of what she already knows and feels.

C: Um-hum.

Act V: Making Yourself a Spiritual Eyeglass Holder

BK: So pull them out and put them on. I'm curious what kind of eyeglass case you would get if you just went out to a store now. Do you have an idea of what that would look like? Maybe it's something you make. Can you make things? ... Can your mother make it for you?

(C and T laugh.)

C: Perhaps.

BK: Do you have any idea, any fantasy? What comes to your mind if you had to think, "What is the perfect eyeglass holder to hold your spiritual eyeglasses?" What would that look like? What color would it be? What design would it have?

C: Um, it's like dark leather.

BK: Um.

C: Um-hum ... I see giraffes on it. Little yellow giraffes with red spots.

BK: Wow. That's just ...

C: Appliqued on it, like sewn in rough.

BK: That's beautiful. Did you know that the oldest tribe in Africa honors the giraffe because they see its long neck as a stretching to be with God?

I am talking about the Kalahari Bushmen as a means of showing how her choice of image further validates the spiritual confirmation she seeks.

C: I didn't know that.

BK: They even dance the giraffe dance as a holy dance.

C: Wow.

BK: Isn't that something?

C: Um-hum.

BK: What happens when they get filled with loving and celebrating God? They get out of the way and get themselves so worked up that they actually feel their body stretch. That's why some of the oldest rock art you'll see in Africa shows human figures with their necks stretched like a giraffe.

C: Hum.

BK: They feel closer to God.

In other words, they seek what she seeks.

C: Hum.

BK: This shows me that she's plugged in.

T: She is.

BK: Wow. You're going to Africa.

C: Yah.

BK: Amazing.

C: Yah.

BK: This is all very wonderful.

C: Yah, yah.

T: It is.

BK: Very wonderful. When you see a giraffe over there, you'll have your case, your eyeglass case.

C: Yah.

BK: It might even be fun to just pull it out at that moment.

(All laugh.)

C: I'll do that.

BK: And you'll remember that ...

C: Yah.

BK: ... this is something very special.

She is imagining using this eyeglass holder in her future outside the session.

C: Yah.

BK: Can you create this? Can you bring this into your life? This leather spiritual eyeglass case with giraffes?

C: Um-hum.

BK: Wonderful. How soon can you do that? How fast can you get mobilized to do it?

C: Thankfully, I'm married to someone who is quick on the draw. I'll share my vision with him, and I'm sure we'll come up with something quickly.

BK: I think it's more important than therapy.

The context of acting with spiritual validity is more important than being inside the problem validity of therapy.

C: Okay.

BK: I think it's more important for you to do this than continue with therapy. I don't think it's therapy you're looking for.

(C laughs.)

BK: I think it's sanctification.

C: Um-hum.

T: Um-hum.

BK: But it's not that you haven't had it. You just haven't been pulling your spiritual eyeglasses out of your case.

The problem wasn't that she needed to receive anything but that she wasn't holding what she had received. She had spiritual seeing but needed something to hold her spiritual eyeglasses.

C: Yes.

BK: And putting them on.

C: Yah …

BK: Very interesting, isn't it?

C: Um-hum.

BK: If you've been wanting this anointment, then it's already been given.

C: Um-hum.

BK: You just haven't been using it.

C: Um-hum.

BK: That eyeglass case is simply a reminder, because we're all human and we forget.

C: Um-hum.

BK: Once sanctified, it's not taken away.

C: Um-hum.

BK: Once you've received, it's always present. People sometimes forget.

C: Um-hum.

BK: She with the spiritual seeing ...

T: Yah.

BK: ... has the world's first sanctified spiritual eyeglass holder.

Marking the moment as unique also marks her as having a unique mission and path in life.

T: Yes, yes.

BK: That's exciting.

T: She has the market.

BK: Wow.

T: I would want to see the case.

C: Um-hum.

BK: When this is in your life, you're not going to have a need for therapy. That's probably going to be disappointing for you because there is something interesting about therapy.

Therapy is a metaphor for thinking about all the choices and paths she can brood over.

C: Um-hum.

BK: Isn't there?

C: Yah.

BK: It allows you to sort of put off the work that you're supposed to do. It's like a time-out.

C: It is a time-out for me.

BK: You have the anointment. You know there is something for you to do, but you may be tempted to say, "Maybe I'll just wait a little while longer. Let me have a little more vacation time." Therapy gives you a little vacation time.

C: That's hard, though.

BK: It's going to be hard to leave your therapist because you'll get tired when you get on the Lord's highway. You'll get weary, but you'll also be exalted and know that this is what you're supposed to be doing. You'll have as many lines of feelings about this as you thought there were many roads. Some roads will be frustrating and some people will try to challenge you. Some people will even quote scripture and say, "You don't know what you're doing." When you hear that, you just need to take out your eyeglass case, act as if you're pulling out some glasses and put them on. No matter what all the differences are, you will see and feel that there is one road. I think you guys should talk about how and when it's time for her to get to work.

The fact that she is human is underscored again, spelled out as meaning that she will always find times when she is confused about seeing different choices and roads. All she needs is to remember that she has spiritual seeing at hand, ready to bring her back in focus.

C: Um-hum.

T: Hum.

BK: Because vacations get to be exhausting, too.

T: Yah.

BK: I know that. Take a vacation and after a while, you can't wait to get back to work because you're feeling exhausted.

T: That's right, I gotta go rest.

BK: Right.

T: Go back to work.

BK: Good. So you're going to do this?

C: Um-hum.

BK: You're making me very happy.

(T laughs.)

BK: Why don't you guys talk about this a bit, and I'll come back and say a few more words a little later on.

C: Okay.

BK: Really a pleasure to meet you.

C: Thank you.

BK: Bless you. It's all my pleasure.

C: Thank you; thank you.

BK: Thanks.

(BK leaves the room.)

T: How was that? Pretty good?

C: Yah.

T: I'm kind of in a trance feeling …

The session was a trance for everyone, as we entered a spiritual frame of reference.

C: Yah, yah … That was pretty prophetic.

T: Ah.

C: Yah.

T: Fascinating, isn't it?

C: Yah.

T: Tell me of your plan now, of making the eyeglass case.

C: I don't know, I …

T: You want to incorporate your husband …

C: Yah. I'll have to get some leather and …

T: Can you make it?

C: Make some leather?

T: Well, that would be a chore. You could do that. You can tan leather. I know you're a pioneer person, but you could also buy a leather case and put giraffes on it.

C: I don't know. I think the eyeglass holder should be a pretty, but rough, little thing.

T: We're rough people.

C: Yah, yah.

T: Born centuries too late, you know?

C: Um-hum.

T: In this modern stuff, I mean. I think it's important to do it.

C: Um-hum.

T: Do it quickly.

C: Okay, I will.

T: I want to see it.

C: Okay.

T: The rougher, the better.

C: Um-hum.

T: That's fascinating.

C: Um-hum. I think we'll start from scratch.

T: So you're gonna harvest an animal?

C: I might.

T: What kind of leather do you want?

(Knock at door)

T: Come in.

(BK enters the room.)

BK: I just got another idea.

T: Yah.

(T and C laugh.)

C: Great.

BK: It'll be all right.

C: I'm going hunting now. I'm not sure that [my therapist] is comfortable with me going hunting for the actual animal that I'm going to skin. He's not sure what to do with that.

(All laugh.)

BK: That's his mold.

T: Yah.

C: He wants me to go to the store. I'm going to the woods and …

BK: Bow and arrow?

T: I couldn't help it.

BK: It's important to remember that of all these different lines, different ways of knowing, include the way of the mind and the way of the heart.

C: Um-hum.

BK: We want to make sure those are aligned. In our culture, people talk about the heart, but they usually set it up so it's a head talking about a heart.

C: Um-hum.

BK: Spiritual presence, I think, is more about a heart holding the head.

T: Yah.

C: Um-hum.

BK: If you … think about choices, you're all in your head and … it's going to be all over the place.

C: Um-hum.

BK: Because the mind's tricky.

C: Um-hum.

BK: It can just think of every kind of reason for this way and that way.

C: Um-hum.

BK: People get confused. But when the heart's right, the head is within the heart. Then things are set in a good way. Maybe as part of helping you remember to bring your heart and mind together, when you take out your spiritual eyeglass case, tap your head and your heart with them.

I am addressing the potential dichotomy of unnecessarily splitting her mind against her body and moving to make them one inseparable unity, discussed as moving from a heart concept in the head to a head organized by the heart, which is implied to be spiritual presence.

C: Um.

BK: As a way to get them lined up as one line.

C: Um-hum.

T: Um-hum.

BK: You know what I'm saying?

C: Um-hum.

BK: Just a little ritual. Pull them out and tap yourself.

C: Um-hum.

BK: Just a reminder of the one line. That line goes straight—all the way up.

I point to "heaven" as I say this, indicating that the one path from her heart is her line to the sacred.

C: Um-hum.

T: Yah.

C: We've talked about the heart. ... I can't get away from my head. I'm so used to telling my heart to be quiet.

BK: If you had your eyeglasses, you'd know what to do right now. You'd be able to take them out and go tap, tap, tap, and then put them on. When you put them on, what do you need to be able to see through their spiritual lenses? You need to trust.

C: Yes.

BK: You have to hand it over.

(T laughs.)

C: Um-hum.

BK: When you take them off, you will grow tired from time to time, and start questioning things. ... That's when you need to get out the glasses.

We are weaving this ritual into her everyday life.

C: Yah, yah.

BK: Remind yourself.

C: Yah.

BK: What sometimes helps is a sacred song that touches your heart.

C: Um-hum.

BK: The most spiritual people I've known in my life line things up by stopping the talking and handing it over to a song. Why? Because a song brings up the heart.

This is what the elders of diverse healing cultures taught me—music centers one in the heart and stills a chattering mind.

C: Um-hum.

BK: You see.

C: Um-hum.

BK: You know how that works.

C: Um-hum.

BK: You need to add music to your lyrics.

C: Hum.

BK: All sacred songs were anointed, brought into the world to help us line things up.

C: Um-hum.

BK: Pull out your eyeglasses, tap your head, tap your heart, put them on, and bring forth a sacred song that fills your heart, soul, and mind.

C: Um-hum.

BK: It's the missing piece. You already have the spiritual eyeglasses. Those were given to you …

C: Yah.

T: Yah.

BK: By the One above. You just need to …

T: Add the case …

BK: That's it. That's all of it. That's the whole thing. That's what people have used for thousands of years and they haven't needed anything more.

C: Yah.

BK: Once they get to the place where you are, they just need to remind themselves, they need to see, hear, feel, and be in the world through spiritual eyes, ears, and knowing.

She came to therapy to remember what she already knew, to be reminded to use what she already has, and to be ritualistically grateful for the greater source that leads her along her blessed path.

C: Um-hum.

BK: Full presence.

C: Um-hum.

BK: We must remind ourselves in some special way that when we see all the confusion, all the lines, we just need to put on spiritual eyes and make the heart and the mind one. This oneness is like the soulful presence of the whole of us.

The one path she was looking for was an integration of heart and mind which, for her, is accomplished through spiritual faith.

T: Ah.

C: Yah, yah.

BK: To facilitate this being full without any distraction, we pour forth a song.

C: Um-hum.

BK: A song stills the mind.

C: Um-hum.

BK: A song brings the feeling—that's the heart.

C: Um-hum.

BK: This room is like the green room of a theatre.

C: Um-hum.

BK: Life is out there waiting for you. We're behind the stage. Your vacation has been like sitting in the green room … Now you've got it. You'll go out that door as an anointed woman with spiritual eyes, someone who knows how to line up your heart and mind with a song. Hallelujah!

Here I reframe therapy as where she has prepared herself to get on the stage of the everyday, fully ready with all the lines and props she needs.

(T laughs.)
(BK shakes C's hand.)

BK: Give me a hug. I know you're going to be, not fine, you're going to be really fine.

(C laughs.)

BK: Okay.
T: Yah.
C: Yah.
BK: You're going to be a fine line. Going to be a fine line full of love.
T: Yah.

(C laughs.)

BK: Okay.
C: That's good.
BK: What a pleasure.
C: That's good. Thank you.
BK: You, too.

(BK leaves the room.)

T: Ah. You ready?
C: Yah.
T: The world awaits, uh?
C: Um-hum.
T: You have your song?
C: Um, I will.
T: Yah.
C: I will. And it will be an amazing moment.
T: Um-huh.
C: Um-huh. Because it will be confirmation …
T: Yah.
C: … of all that we've said.
T: You can find it. You can always find it … You did good today.
C: Well, I'm the blessed one (laughs).

T: Ah, we are all blessed.

C: Um-hum.

T: We are blessed for you being here as well, thank you.

C: Was that encouraging confirmation to you?

T: Ah yah.

C: To know that what you've been speaking to me is exactly what he spoke about today.

The client is bringing everyone into line, all contributors acknowledged, congruently coordinated, and sanctified, thereby bringing forth what she had requested.

T: It's what the Great Spirit speaks.

C: Exactly.

T: Ah.

C: To know that you've been participating in that.

T: Very nice, a nice feeling.

C: But to have that confirmed.

T: So you're ready.

C: Yah.

T: Let's go.

C: I've got to go hunting.

T: Yes. Absolutely. I hope so.

C: Actually, I know hunters and I might see what pelts they'll share with me.

T: Good for you.

With invisible eyeglasses this young woman went to Africa with her husband, seeing and feeling its untame splendor and complexity. You can correctly assume that she saw giraffes in a way that most people would never imagine. Upon her return home, she had a way to keep her mission in focus and a means of lining things up. She has fond memories of this session, as she later reported, and said it was a special holy moment in her life.

CHAPTER 11

Making the News

Sam (C) is a middle-aged war veteran who has been diagnosed with post-traumatic stress disorder (PTSD). Two student trainee co-therapists (T) at a university clinic begin this session while I observe from behind a one-way mirror. Sam, who holds a book he brought with him, initially launches into a speech about his situation.

Act I: PTSD

"I am overwhelmed by the alienation I've received from everybody I know. I've never had fewer people to talk to in my entire life than I have right now. … I can't help but wonder how many other people in the area are in the same boat. I thought what I want to do is start some sort of a support group. I just can't believe I'm the only one who's in this sort of situation with no one to talk to. It's absolutely isolating, alienating, and mentally terrible."

He goes on to say that he is in a gloomy mood. "I'm waiting for something good to happen, and I just can't seem to turn the coin and have anything come up positive and I'm just tired of it. I'm disgusted with it. … I've never dealt with this stigma before, but boy, I got one going on now. Like a chicken on a junebug it latched onto me. Everything and everybody is categorized, identified, separated, and stamped. … Everybody and their brother has had a conversation with somebody else and their brother about me in particular; and no one's talked to me!"

"The thing that shocks me the most is the pretense that so many people in this area use to guide and govern their lives—their pretense about church and religion. … I'll buy it, but I don't want to hear any more about it. Don't tell me about the words written in red. Christ was here for a purpose … to teach and to minister and … he left very specific examples to go by … Christ's existence on earth is to straighten out everybody else. … I'm just sick of all these good Christian folk and their attitude about what's going on in my life. It's like, man, I'm just tired of being alone by myself. I'm tired of having to want to talk about it all the time. I'm tired. I'm tired of it to be quite honest with you."

"I realize that it's going to be a methodical, calculated approach … and each step of the way will have to be judged and critiqued to see if you want to continue along that venue or whether you want to switch paths and go in a different direction … I'm talking about my particular situation in general … Here's what I think needs to happen. … You know my approach to statistical process control. I like to think everything out and put it into perspective."

"I have a strong medical background and for years before I got back into manual labor activities, I did a lot of consulting with accreditation and compliance issues and things. I would like to see a structured care plan in a flowchart format that says, okay, here's the deal. Here's Sam, here's his therapist, here's his doctor, and then each component of that trilogy or whatever it's going to be, have their responsibilities manifested and put down on paper so I can track my progress. Now we've come to the point in the therapy—and I'm sure there's some form of therapy plan—where I'd like to see this generated in a format so I can see that I'm making progress. … Did it work? If yes, then go on to this step. If no, then go to that step. I want to have some way to guide me through the process. Now if I have to develop that concept of logic myself, we can. May I?"

Sam then stands up and arranges a board with poster paper attached to it. With marking pens, he starts drawing some diagrams, pointing out where he and the therapist are located in the diagrammed scheme. He delineates the sequence of how treatment should proceed forward, advising "everything we do involves process." He spells this out further: "There is a process to the way you pick your nose (the therapists laugh), the way you make a peanut butter and jelly sandwich, and the way you fix physical issues. (Sam then points to the board.) This is a representation of a written care plan. … The therapists [should] evaluate all the input and give me feedback and the feedback at this particular point is that I have PTSD. At this point it's like a big question … now a couple things have to happen … someone's got to educate me."

Sam proceeds to describe how he has learned about the "trigger points" that can activate his symptoms, saying he learned this from his therapists and some self-help books. He goes on about this education and concludes: "The most important thing you can have as you approach your PTSD issues is the proper support group in place."

Sam raises his voice and then pronounces loudly and dramatically, "I can't get anybody to listen to me. I got to come over here and pay 10 dollars a pop to have two strangers that have no world responsibilities or affiliations with me, in order to hear every one of the issues that concerns me. That in itself is okay, but where is that in the process? It is at the point where I don't have a support group. I don't have anybody that wants to talk to me. I don't have anybody. My wife, ex-wives, brothers, sisters, cousins, aunts, uncles, kids, and their kids: NO ONE wants to touch me. 'Oooo, Stay away from him, he's got the PTSD.' PTSD is transmittable. There's no inoculation for it. It's worse than herpes virus simplex. … I can't believe that there's not such a thing as a care plan or a therapy plan when you come into mental health issues. If it's not there, it should be. Based on the information here (Sam leafs through a book he is holding that is about PTSD), I could write a therapy care plan for myself. … Right now, it's like my ass is on fire and my head's catching and the faster you run through the woods and scream for help, the faster the fire burns and the hotter it gets."

"I want to formulate a care plan in my mind. … And I could care less about religion. I could care less about the symbology. It's the man behind the cross. That's what makes the difference to me. Everybody professes all these Christ-like principles and beliefs and faith, and it's like, where are their examples? Show me an example. Show me with me! I do it. If I see a guy beside the road changing a flat tire, I'd stop to help him. If I see a guy walking along the road with a gas can … I'd stop and take him to a gas station."

"I had a brother who came back with post traumatic stress from Vietnam. … I didn't do anything. He just went crazy and Daddy called me one night while he was throwing things and breaking stuff. Do you think I'm going to sit around and listen to these stories chosen from the words written in red and … dissect them and bisect them, laying them out in front of you? … That's not what Christ wants. God wants us all home again to live and love one another like there's no tomorrow. … I want to see the validity of that 2,000-year-old concept applied in my life right now and the lives of all those people around me. Okay?"

"I don't want to be the only one making suffering and giving gifts anymore. Right now I need more than I have the capability to give … God, please send somebody to help me heal. Every day … it's like I turn today's newspaper page and say, 'Wow, it [something terrible] happened again.' Then I turn another page tomorrow and it's like, 'Man, where did this [terrible event] come from?' When am I going to turn the newspaper page and find a different story? Why can't I find a story to share with my kids?"

"My own 18-year-old daughter says to me, 'Daddy I can't come home right now. You're in this process of trying to heal and stuff. I can't be around you and go to school and work full-time while you're not a hundred percent right … I keep trying to tell her that I read and I learn and go through therapy … Or I go into the woods to think and. …"

(There is a knock at the door.)

T: Come on in.

(Brad Keeney [BK] enters the room.)

BK: Hi.
C: Hi there. You come in to referee?
BK: I'm Brad Keeney.
C: Hi, Brad. (BK and C shake hands. C then looks at the therapists.) Looks like Jesus, don't he?
BK: (Sitting down) I was just going to say that when it comes to your situation, you are one observant man. You have a highly skilled talent for observation.

His whole previous performance of graphing his situation and explaining how therapy should be organized with detail about what he observes is reframed as a skill.

C: But you know what it's worth?
BK: Don't—we don't have to go there yet.

Before he can disqualify or downplay his skill, I will utilize his question about its worth as a way of underscoring how his skill has a practical cost or consequence.

C: Okay.
BK: The consequence of this skill is that you notice more than other people notice. And what you notice is sometimes disturbing.
C: It affects me.
BK: That's right. What I want to say to you is that there's a cost to having the skill of seeing so much. There's a cost to noticing every day that it is the same newspaper story. Every day it's the same disappointment. What should happen doesn't happen.
C: It's been the same for 2,000 years.

Act II: Experiment: Making the News

BK: It's been that way for a long time. What I want to suggest is that maybe it is possible that no one has ever really stepped up to the plate to do what should be done. If that's the case, then you've been waiting for a long time for somebody to do it. It just isn't going to happen. Now, please go with me.
C: Okay.

BK: 'Cause you're seeing the same story every day: same story, same story, same story. Every day you look in the world and you ask, 'Why isn't somebody enacting and embodying the red letters as opposed to just putting the letters out in front of you?'

C: He [Jesus] wasn't a complicated man.

BK: I've got an idea. I just want to hear what you say about it. What if, as an experiment—because the only way you're going to get relevant feedback is to experiment—you try something in order to see its outcome. Depending on the outcome, you'll shape it one way or another. … See if you can determine whether the feedback's going to be negative or positive. You have to put a difference in the world as an experiment to see whether you can notice whether it goes one way or the other. What if you did something that you're waiting to see others do? It could be small. It could be just saying one little sentence to someone that constitutes the thing that should be said. It could be one small action. It might not have to be toward a human being. It could be doing something for a sparrow. It could be for an animal you never meet when you go into the woods, maybe leaving it a little seed. Do one small thing and then write it up like you were a newspaper reporter at school: 'A man, unidentified, without anyone asking him to do so, committed this act of goodness in the world today at 11:05 a.m.' After you write it down, scotch tape it in a newspaper and leave it in different places so people would, for once, have an experience they never had like you've never had. 'Cause I guess there's other people in the world just like you, but they are just not as observant … They're reading the newspaper and rather than being disturbed in the way you are, they're disturbed without knowing why they're disturbed, which makes them really more crazy. Do you hear where I'm going?

I am using all the resources he mentioned before I came into the session, doing so to call forth his action. He is now being invited to be the performer, observer, reporter, and news distributor. He is literally invited to make the news.

C: Yeah, I do.

BK: What if you did that? 'Cause I think if you did that, you might be one of the first. I don't know if it's ever been done. In fact, I'm inspired to say this: I've never seen someone come into a clinical office and teach like you taught and observed like you observe. I thought to myself, 'You're inspiring me to encourage you to do something else no one's probably ever done before.' Which is to do anything that's an enactment of being good. Anything without thinking about it ...

I have never seen a client pull out a large sheet of paper, diagram a systems process control analysis of therapy, and then ask for new action and direction. I frame this as a unique event that, in turn, justifies my asking him to do other things that are unique and different. In this case, it is conducting an experiment that requires him to perform the very conduct that he complains about not seeing anyone else do, to see whether he can notice and further implement its effects and organization in the social domain, discussing it in the same way he is able to diagram the feedback process in anything from handling a nose itch to making a sandwich or laying out a sequence of action for therapy.

C: Right.

BK: Choose anything you can do, do it in the world, write it up, and then scotch tape it into a newspaper and leave that newspaper for someone, whether it's for your family, someone in a library, or somebody at a café. ... They will open up the page—maybe put it on page two because most people get to page two. Or maybe put it in the sporting news; a lot of people go right there. Put a little article about something you do. It's going to be a report about an anonymous man, a person who asked to remain anonymous, who committed an act of kindness today at 11:05 a.m., or whenever you happened to do it. This, I believe, is something that will start a new kind of loop, a new kind of action that will show you what happens if others see what you're not seeing others do. That's all. (BK stands up.) Just talk about this for a while. I just want you to think about this. (BK puts his hand on C's shoulder.) I think if anybody can do it, you can do it. I know it's a thing you weren't expecting to hear. ... (BK looks at the book in Sam's lap.) What's the name of this book?

The prescribed task enables Sam to see what he's "not seeing others do." As he has taught the therapists how to sequence movement toward a solution, he is asked to enact a sequence in the area of his concern that lies outside the clinic. The implication is that he will be teaching others how to do it. I point to the title of the book he is holding as a way of enacting this change in the here and now of the session.

C: *I Can't Get Over It*. It's very informative. Very enlightening, ah ...

BK: Do you want to get over it?

C: I, I, I'd love to.

BK: Then the first thing you should do is just cross out that title. Just say I
can get over it. And then tonight before you go to sleep, imagine
what would be inside a book with this new title. I'm serious.

I am trying to start a self-corrective loop for his relationship with the
book that explains his diagnosed condition by changing the title from
"can't" to "can" so he is in a different relationship with the book and can
start pondering the implications of such a shift in meaning.

C: Okay.

BK: These loops you talk about—you've been going through a lot of loops
that go round and round and round and round. You're waiting for something to happen that isn't happening because you're
waiting to see it performed by other people and they're not performing it. Maybe they haven't for 2,000 years. Your experiment
is to see an act of kindness and an act of goodness, no matter
how small, and then report it. I don't care if your news report is
two or three sentences or one, two, or three paragraphs. Write it
up. Either by pencil, pen, or typing—whatever you wish. Scotch
tape it onto a newspaper, on page two, and leave it for someone to read. Start something that has never appeared before
your eyes from the performance of someone else. You might
just be the beginning of a snowball that gets bigger and bigger.
Because if one person reads that story, I believe they'll recognize it in the way you would recognize it and say, "My heavens, I've never read a story like that in the newspaper." It finally
happened. After 2,000 years, someone actually did something
that is newsworthy in a positive way that enacts and embodies
the red-letter words. No matter how small, because you know
it doesn't matter how small. Because goodness is not measured
by saying that it's big or good. Sometimes, the smaller it is, the
more important it is. But that doesn't matter. It can be as simple
as taking a pan of seeds and putting them in the middle of the
woods just for a sparrow, if that comes from the goodness of
your heart and wanting to create a good deed. If a man did that
out of goodness and good intention and wrote about it for the
newspaper so that someone else could be inspired, the world
might change. And you then will know that someone has seen
what you've been able to see and that's a new kind of loop, a new
kind of feedback. Now focus and don't let anything drift. Stay
right on this. Let's stay in that loop, and I'll be back in a few
seconds. Okay?

(BK leaves the room.)

I have restated the experiment so that it holds many of the resourceful frames he has presented. This shift in the theme of the session will be allowed to settle in. I leave so that the therapists and client can get more used to being in this altered place.

C: I, I, I do that. I do that on a regular basis. I believe the most important thing you can do is, is live the old golden rule, but, but ...

T: What's the difference between what you've been doing and what he's described.

C: I've not been recording it for, for, anybody ... I've been trying to live by example.

Sam indicates that he is already in step with this experimental prescription. The missing piece is the news about his good deeds. The action has already begun but hasn't been reported in the newspapers.

T: So you haven't been looking for the feedback?

C: No. I haven't been.

T: Okay. Because feedback isn't necessarily what we get back, it's the change that takes place.

C: I, I don't ... I was, I was a creature estranged from God in a multitude of ways. The estrangements were associated with my condition. I had no idea, but I reached a pinnacle of lowness and I said to God, "God, that's it. I can't go no more." I want to forsake all of what haunts me because I, if nothing else, I got to live guilt-free for a while. So I'm not going to smoke, drink, do drugs, chase women, and all those kind of things that, you know, add to complexity, and, and, and I've always been an exceptionally giving individual. I'll give you the shirt off my back. I do it anonymously as much as I can. But a lot of my giving is to get that recognition, to get that attention, it's to get that love I never knew when I needed love.

I hear Sam declare that he has addressed God and has indicated he is willing to make sacrifices. I hold on to this as something I can utilize as a resource.

T: Right, but this is different in the aspect that you are doing it anonymously but at the same time you're putting it out there for others to see so that there's this chain reaction ...

C: I, I ...

T: There's this movement that you create, this new cycle. This new perturbation that you see is new feedback. And not just in your life. 'Cause that's what you're talking about: change that is not just about Sam, but change that affects the world. It can strike you … but when you touch one domino, it eventually touches another.

C: So it's like asking me to pen a new parable.

T: Yes.

C: You know, you live a certain way and you create a certain environment and then you pen the parable and pass it on.

(There is a knock at the door.)

T: Come in.

(BK enters the room.)

Act III: The Mission: Be the Good News (and Change the Title of Your Book!)

BK: I wanted to leave the room last time because I felt a powerful opportunity in this room for you. (BK looks directly at C.) It's clear to me that over the years you've been in some ways—sometimes knowing it, sometimes not knowing it—really asking God to tell you what your orders are.

This communication is based on the fact that he previously mentioned his communication with God. As a former military man in therapy due to a combat-related challenge, it follows that he awaits orders from the superior commanding office who is in charge of his life.

C: YES!

BK: What am I supposed to do?

C: YES! (BK shakes C's hand.)

BK: You try to show God, "I will make the sacrifices. I will even try the things I'm not supposed to do in order to know how much it costs to give more."

I am simply recycling what he has said, doing so in a manner that moves us toward a new contextualization of his life.

C: YES! (BK shakes C's hand again.)

BK: So I'm suggesting that it may be possible that in 2,000 years no one has committed the small acts of beauty and of goodness and of kindness. You have been observing how this isn't happening. But it takes a teacher to show the others. I saw a man here teaching. Wow. I thought and I actually joked behind the mirror saying, "I should get college credit for watching this teacher teach you two as he very clearly drew out the way in which a process should be outlined in a schematic way …"

His performed teaching of the therapists in this session sets up the idea that his mission is to teach in general.

C: YES! (C enthusiastically nods several times.)

BK: "... drawn so that it's showing that we're here for a purpose, we're here for a mission, and we must know where we aren't." I believe that maybe what you see others not doing indicates your mission: to show them. That's what I want to ask you to do this week. At least once, maybe twice, maybe three times, maybe every day, but at least once, perform what others haven't done, a small act of goodness or kindness that is then written about by you, put in a newspaper, and then left for someone to read. Can you do that?

The pivotal point here is saying that what he does not see in others is the way he is being shown what he is supposed to do. This connotes that which disturbs him as the answer he is looking for: a message about the mission he should pursue.

C: Sure.

BK: Because that may be your mission. And if it's not, it's the first step towards finding your mission. That's more than being in your mind trying to sort it out. Because you can't get out of the first box [of his diagram] to the next and go around the loop until you step into the world and do something to see how it has an effect. We call that, as you know, "feedback." But it doesn't matter what you call it. It's just your time to do it. And I would say that the first time you do it, come home (BK puts his hand on the book in C's lap) and scratch out the letter "T" in the book's title, so it says, "I Can Get Over It." Will you do that?

Whether this new action is his whole life mission or not is not presented for discussion, something that can easily lead to an irresolvable debate. Rather, it is defined as pointing toward the process that brings it forth—a step toward realizing his goal.

C: Yeah.

BK: (BK shakes C's hand.) Congratulations. You just may be somebody who starts to get the world moving in a good way.

C: Well, it's a scary thought.

BK: Amazing, huh? (BK looks at the therapists.) There's nothing more to understand here. I think you should celebrate with him over what he's going to go into the world and do. Talk about that being a good thing and just honor this moment. You might even be silent just to give it some sacred respect. I'm moved and I'm touched. I see you're a man of depth and you have very powerful longings to be good in the world and to see the world move in the way that it should be moving, should have been moving for many, many, many years. Okay? So it's time for you to move so the rest of the world can move with you. I thank you for teaching us all through your commitment to having fully waited to find the answer to, "What is my mission? What am I to do?" You have shown your worthiness. But you know it's what you see in the world that's not done that tells you what needs to be done by you. That's how we're sort of put together. Because when we notice what isn't being done, it tells us in some way, "That's what I need to do."

Do some small thing. I don't know if it's going to be in the woods, or in the house, the garage, or on a sidewalk. I don't know. That will be for you to decide. Maybe you already know. If you do know, just don't tell me about it; write about it. You can use several newspapers. You could write it up four or five times and leave it reported in four or five different places. Wouldn't that be something? In three months there might be three, four, five hundred, six hundred, maybe a thousand people saying to one another, "I got a newspaper with this article." Another says, "I found one with this article. It seems like they're all positive stories." (BK looks at C.) You see how powerful that is? (C nods yes.) I'm really happy I came up here to meet you. (C nods again in agreement.)

(To the therapists and C) I think you should celebrate— maybe it's a silent prayer, maybe it's a silent respect—just to thank him for having come here to teach us these ways and what he's capable of doing. Okay?

Because he easily gets burdened by his excessive thinking and talking, I am prescribing silence as a way of honoring, respecting, and celebrating his found mission in life. It is time for him to bring forth the good news by being it, reporting it, and circulating it.

T: (Looking at C) There is something I think you've been asking for over a long time. You said you just want somebody to embrace. We can embrace at this moment.

BK: It's up to you. But don't forget this part. (BK points to the book on C's lap) When you place that first news report in the world, take that "Can't" and make it a "Can." It's very important. Because that means you've changed the name of the game. You've changed the name on the book, you've changed the name of your life, and in so doing, you've changed the possibility that others may go in the direction they always should have but have been stuck, because it took someone committing one sacred act of embodying the red-letter words, not in a grandiose way but in a simple— ambiguous, perhaps not ambiguous—way. It's whatever comes up for you. Okay? (C nods his confirmation.)

(Looking at the therapists) When you say, "embrace," what do you mean? To celebrate? (The therapists stand up to hug C.) That's good. Good. That's a good moment. It's a moment that we'll all remember. I might get in there, too. (BK hugs C.) It's good, man. It's been a long hard road to get yourself prepared, but the longer the preparation, the greater the curing of the whole being.

Changing the book title continues to be a metaphor for the change he seeks. I underscore this saying it changes the name of the game and the name of his life as well as contributing to changing the lives of others. His long road of previous suffering is now, in this contextual home, given the meaning of being prepared by life to find and be able to accomplish his mission.

T: Hey Sam. (The other therapist hugs C.)
C: I appreciate it.
BK: You bet … it's waiting for you, man.
C: That's a scary thought.
BK: It's all right. 'Cause I think a child can change a world that's not getting somewhere.

My comment was based on the scriptural reference that suggests that, unless you become like a child, you can never enter the kingdom. This helps keep the final scene in the spiritual domain that has deep significance for the client.

C: Christ wasn't a complicated man.
BK: You got it.
C: He didn't come to live with complicated people. He came to live with the simple man, the simple teachings, and …
T: You got it.
C: … and we just seem to repeat all of the same issues over and over again. And somewhere it's got to stop. It's got to stop for me.
T: You become the living parable. I think that's what you're talking about.
BK: That's a nice way of putting it. Okay. I expect that "T" to drop off of that book.

C: All right. We'll see what happens.
BK: Excellent.
C: Let's get out of here. I appreciate it.
BK: You bet.

Sam reported that he was able to move forward with his life without the need for any more therapy. He said he was going to start a support group and use what he had learned to help others who had gone through what he had experienced. He believed that he had come closer to God through his time with us. He was encouraged to keep enacting and reporting the "Good News" because the newspapers needed that.

CHAPTER 12

Magmore

This session involves the son, Andy, a 16-year-old boy from Louisiana (S); his grandmother (G); and his "mildly mentally handicapped" mother (M). A year before, he had threatened to kill his mom with a steak knife because she refused to play with him. Andy told the sheriff's deputy that he had problems and wanted to get some help. He had been in and out of hospitals and institutions since he was 7. Andy had a history of torturing, mutilating, and sometimes being sexually inappropriate with several different types of animals, including cats, dogs, frogs, birds, and lizards, which he would also roast and eat. Finally, he saw ghosts when he was admitted to a hospital. As I walked to my clinical office where they were waiting with their therapist (T), I noticed a soft rubber ball on the ground. I picked it up and carried the ball with me as I entered the room. (Please note that this session is included on track 2 of the DVD.)

Act I: "Not Really"

Brad Keeney (BK): How y'all doin'?
T: Good.
BK: (To T) Is this your ball?
T: Yeah, I think I had it on my desk.
BK: Did you lose it?
T: I did.
BK: It was just sitting outside.
T: Well, I guess it's yours now.
BK: It looks hard, doesn't it?
T: Yeah, it does.

BK: It looks hard.

T: But when you touch [squeeze] it, it's soft.

BK: (To C) Do you think it looks hard? But feel it: it's real soft. See? (BK throws ball to C.)

S: Yeah. (Andy throws ball back to BK.)

BK: Wow. So is that the way you are? Do you look hard, but you're really soft?

I simply utilized what I picked up on the ground before entering the session. The ball did look hard and it surprised me how soft it felt when it was squeezed. I was thinking about this as I walked into my office, where the family was already waiting. My first impression of Andy was that he looked a lot softer than the hard descriptions some outside professionals had made about him.

S: Yeah.

BK: (Laughs) Tell me why are you here? Why are you guys here? (G laughs.) What's the reason?

G: To meet you for one thing, and to talk, chat for a little while.

M: Yeah.

BK: What kind of creative expression led all this to take place today? What happened? Did he burn a building down or did he …?

I sometimes avoid saying "problem" (but not always) and use metaphors like "creative expression," "surprise," "trick," or "unexpected performance." For example, "What kind of surprise did you cook up that has everyone talking?"; "Did you go trick or treating when it wasn't Halloween?"; or "What unexpected performance did you star in that has your parents confused?" These questions already plant the seeds for moving us to a more resourceful context for holding the experience.

G: He tried (laughs).

M: No, I did.

S: No, I did.

BK: Hold it, hold it …

M: I did.

S: She accidentally …

M: … I accidentally left the grease on the stove and I forgot to turn it off and then smoke went all over the house.

BK: So mom tried to burn down a building.

M: Yeah, yeah.

T: … the house.

M: That's me, that's me.

BK: So mom's on fire, mom's a fire setter …

M: Yeah … Not really.

BK: Grandma, what do you do to make the family interesting? Do you flood the house?

(Family laughs.)

T: A hurricane will do that, huh?

G: Yeah.

BK: What makes you interesting? What makes people think, "Yup, you belong to this family?" … I should say, that these people belong to you?

We have quickly moved to including all family members in the definition of the situation. Here so-called "problems" are set within the frame of that which "makes the family interesting," a softer, more pliable frame that makes it easier to get out of than any concretized, hard conceptual walls or framings like psychopathology.

G: Well, I can't say that I gripe all the time, but sometimes I do.

T: Yeah.

G: I take Paxil and they say when I don't take my Paxil …

M: … Oh yeah, when she don't take her Paxil, oh my God!

G: … I'm grouchy (laughs).

M: Yup, you can't be around her.

BK: (To G) So you gripe all the time.

G: Not really.

BK: (To M) Sometimes you set fires, but not really. (To Andy) And what do they say about you that's "not really"?

I have observed that both mother and grandmother follow their comments about what could be taken as a problem with the phrase, "not really." This gives me a useful distinction to address Andy. Namely, he is a member of a family where people are "not really" what they seem. Just like the ball that looked hard; it was really soft.

S: Oh, I just aggravate people.

M: Yeah, he just aggravates his mama.

BK: But not really.

S: No, not really. I just play a lot.

BK: (To T) So, these people do things, but not really. You know what that tells me? That tells me that whatever he got in trouble for, he's just pretending.

The "not really" modifier enables us to split the problem frames that the family has lived with—dividing them between the distinction of appearance ("not real") and what is not seen ("real"). Each family member speaks this way. I am speaking their language but am cognizant of how it provides a possible way out of their stuck situation. If their surface problems are not what they seem, then let's explore where other alternative meanings might take us.

M: Yeah, he ...
BK: That right?
G: Well, actually he ... (laughs)
M: He wrote on a wall.
G: ... let me see, how am I gonna say this ...
S: ... I spray painted a wall.
M: Yeah, there ya go.
G: He spray painted ...
BK: You spray painted a wall?
S: Yeah.
M: His door.
BK: Hold it. Hold it. Did he do a good job?

I always look for what is resourceful, useful, creative, or transformative within any performance, expression, or account. More generally, I am looking for the implied other (opposite) side of a distinction. For example, if another family with several children complained about one child, I'd ask about the difficulties associated with whoever is the well-behaved sibling: "There's a down side to always being perfect. What are you missing out of life when you devote yourself to always being the perfect child?" In Andy's particular instance, I focus the mention of his drawing on a door (assumed to be bad) on his ability to draw (something good).

G: No, he wrote his name on the side of the building.
BK: Yeah, but was it a good job? Did he not do it in a nice way?
G: Yeah ...
BK: Can you paint well with a spray can?

(Andy nods.)

BK: Really?
M: He put something on his door. What did he put?
G: His initials.
M: His initials.
BK: Does it look nice?
M: No!
S: That's what they think.
BK: Would your friends think it looks nice?
S: Oh, they haven't seen it.
BK: You know what I'm saying? Some people spray paint, but they're really artists.

M: In case your friend did see it …

BK: Okay, so you spray paint walls.

S: Yeah.

M: He does.

Act II: An Aspiring Artist

BK: Maybe he just wants to take an art class.

M: Yeah, he does.

"Drawing on his door" is not allowed to solely take us down the communicative path of his being a "bad boy." As previously mentioned, it also suggests that he wants to be more involved in art. The latter points the way to creative transformation—we are now interacting with an aspiring artist (Act II) rather than the frozen immobility that stems from being perceived as bad (social entry into Act I).

S: I'm a good drawer.

T: He is.

S: I draw creatures.

M: He does.

BK: Interesting. You notice he said the word "creatures." Most kids would
    say "animals."

T: Hmm.

BK: Or maybe you mean creatures, but something that's not an animal?

S: Yeah.

BK: Like what?

S: Monsters.

BK: Monsters? Wow, like what kind?

S: Aliens … like they evolve into a monster and stuff.

BK: Now this is a serious question. Do you do the best you possibly can
    do …

This question suggests that I am more concerned with his doing the best he possibly can with his art rather than where or what he draws.

S: No.

BK: … to draw the scariest monster as opposed to just a monster?

M: He can do that …

BK: No, I want to hear it from him.

S: Not really, but I try my best sometimes.

BK: To make it scary?

S: Yeah.

BK: But, did you notice what you said after you said it? "Not really."
    That's the most interesting thing I've learned about this family.

Anything they say that's about a behavior that others would think is a problem, they always say, "not really." This tells me that they're actually experimenting with various forms of creative expression. In other words, it appears [that mother is] setting the house on fire, but not really. It appears that [grandmother's] complaining and that she's whining, but not really. It appears that Andy's expressing scary things to people, but not really. Does that capture you all?

(G, M, and Andy enthusiastically agree.)

Because the family defined Andy as someone who wants to study art, we have another way to address their theme of "not really." What people think they see on the surface of Andy's scary behavior is not really what it seems. Now that we are in Act II, what is seen on the outside may reflect something that is really about trying to bring forth artistic or creative expression.

BK: There you go.

M: And he scares me when I'm coming out of the bathroom.

BK: See, the problem is the world thinks it's for real. It's really not real ...

M: ... not real.

G: ... not real.

BK: ... that's probably what's happened with your boy. People are seeing something that they think, "Oh, this is real; his monsters are real." Maybe you draw so well that people are scared that you'll actually make a real monster.

S: I'm thinking about it ...

BK: Really?

S ... creatures and stuff. That'd be kind of cool, though, like a new species.

BK: Wow! Now that's what I call ambition! Most kids want to be a musician or be a professional sports player. He wants to create a new species! (BK shakes hands with Andy.)

(Everyone laughs.)

I use every moment to underscore his creativity, praising his ability to think about what other kids don't imagine. This line of inquiry fits the contextual home of a young person aspiring to be an artist.

T: That's pretty impressive.

BK: That's awesome! That takes balls. (Everyone laughs.) That's unbelievable ... that's just unbelievable! You know I would have been so proud if my son would have come to me and said, "You know, dad, I think my goal is to create a new species."

T: That would be pretty cool, huh?

BK: That would be amazing. ... What would the new species ... do that human beings are not able to do? Or, before we get there, what would be something it would say to people that human beings don't say to people?

The new creature, or different species, has become a metaphor that enables us to talk about how to make different, unique, creative expression in the world, which implicitly includes Andy's exploration of various modes of expression.

S: I don't know. I haven't thought about that yet. Like do more. Like they got powers and stuff. Like they can fly ... walk and speak ... talk to creatures.

G: Like that haunted show the other night?

S: I like scary stuff.

BK: So [it can] walk, talk, fly ...

G: He attacks and does whatever.

M: And he watches this movie with bats on it.

BK: Okay, you're just beginning to think about what it would be if you could create it ...

S: What it would be?

BK: ... and what it would express. I mean you're saying it could fly and it could talk and walk. The only thing that's really interesting and different there is that it would fly, unless it talked and said things that human beings don't say.

S: Speaking a new language.

BK: There we go.

T: Hmm.

BK: Exactly. Would it be a language people hear? Or would it be a language where you just put thoughts in people's mind?

S: Put thoughts in people's minds.

The new metaphor provides a way of addressing what Andy's communications have already done to his world—putting thoughts into people's minds. These thoughts, of course, may not really be what they seem.

BK: Would it be thoughts that startle them as opposed to thoughts they're familiar with?

S: Startle them.

BK: Would it be thoughts that you would say they think are real, but you would say are not real?

S: Not really. Yeah, sort of. Yeah.

BK: Okay, so in other words, your family is teaching you how to create the perfect monster.

S: They don't want me to, but I ...

BK: Do you hear what I'm saying, because his monster is going to create fires, but not really. It's going to get everybody to complain, but not really. 'Cause look at Grandma; she's smiling. And she's enjoying this whole family. This family is entertaining, isn't it? Would you say this is an entertaining family? In other words, you don't get bored in this family.

The "not really" family, in other words, can be positively framed as inspiring an aspiring artist to consider all the startling things one could invent to put in people's minds that are not really what they appear to be. Not only does it serve the development of creativity, it makes home life entertaining and not boring!

S: Oh no.

(M and G laugh.)

S: Not really.

BK: Do you know the most complimentary thing that an adolescent, a young man, could say about his family is that, "My family is not boring"?

T: Wow.

BK: Because everyone thinks, at his age, that their family and family life— and that includes him—is boring.

T: That's true.

BK: This is a miracle! You may be the only family that isn't bored in this … state. I don't know. Most families are …

M: Bored.

BK: They're bored. But you have a family in which, my heavens, your son wants to create a new species. Something that doesn't even have to say a word … [because it can] put thoughts into … minds …

M: Oh God, please …

BK: … about things that startle them that really aren't what they seem to be. Very interesting, very interesting.

M: It is.

Act III: Magmore

BK: Let's say you created this creature and did these things. What would be its name?

S: Kind of scary. Something scary.

BK: What was that?

T: Something scary.

BK: Something scary? So we call it SS: "Something Scary"?

S: No, I think um … (puts hands on head)

BK: He's creating right now.

S: Magmore.

BK: Magmore. Magmore or Madmore?

S: Magmore.

BK: That's a very interesting name. I never heard of a Magmore.

M: No, me either.

T: Sounds pretty intense.

BK: It does. Mag is sort of like it's a magnet, so I think people are going to be attracted to it.

T: Drawn to it.

BK: Drawn to it.

T: Yeah.

BK: Does that make sense? (Andy nods his head, indicating "yes.") And the more they try to stop thinking about it, the more they think about it.

S: Yeah … it hypnotizes you.

BK: It hypnotizes. You know what? I've got a feeling that you hypnotize a lot of people. (Everyone laughs.) And they do all kinds of things (laughing) that they think are real that you know are not real.

The name of his monster, Magmore, suggests a magnetic, hypnotic effect on people that leads others to assume more about a situation than meets its real truth. In other words, Magmore embodies the same influence Andy has on the world. We will make this connection more obvious so that any talk about Magmore is also talk about Andy.

M: Oh yeah. He can.

BK: So you're a master hypnotist in some ways.

G: A what?

BK: He's a hypnotist. He hypnotizes, makes people think things that they think are real, but they're not real.

M: But they're not real.

G: Oh, okay.

BK: Like I bet he could convince someone that he actually is trying to make a monster and they would probably send reporters to the house to sneak in …

M: (Nodding her head) … and try to check it out.

BK: (Nodding) … and take photographs and check it out. He could probably convince them of that.

M: I think he can, too.

Even the imagined creation of Magmore is seen as having a likely hypnotic effect on people in Andy's world. Some might actually believe he was really trying to make this monster. That's how hypnotic Andy's influence is on others. He can stir their imagination to believe it is real.

Act IV: Comic Book Artist: To Be (a Cartoonist) or
Not to Be (a Cartoon)? That is the Question

BK: (Looking directly at Andy) Is that right? (Andy nods in agreement.)
   See! Very interesting. Maybe you'll grow up to be a Hollywood
   filmmaker. (Andy laughs) That's the kind of mind. ...

S: I've been thinking about making comic books.

BK: Comic books? There you go: drawing and this kind of creative thinking. Have you made a comic book before?

Andy has led us to another scene, where his specific form of creative
expression may involve making comic books.

S: One time when I was little.

BK: Really? How many years ago?

S: When I was in the sixth grade.

BK: Where is it? Do you still have it? (Andy shakes his head, indicating
   "no.") What happened to it?

S: I gave it to somebody, a little kid.

BK: My heavens. I think this case is done. (BK is looking at therapist while
   Andy lifts his head to look at BK.) He knows precisely what he's
   to do with his life. It's just been on hold. Once upon a time, he
   had a skill that he knew—he recognized he could draw. He knew
   how to use his imagination. He knew how to make things appear
   real (looking at M) that aren't real. That's a comic book. He knew
   how those things that were real and not real sometimes involved
   monsters. That's a comic book. He knew that those monsters
   could set fires but they weren't really fires. He knew that all the
   people and characters in there would be all astir and all complaining and all scared and all thinking this is gonna happen and they
   should be worried, but it wasn't real. And he knew how to create a name for the character that nobody had ever heard before
   because the secret, the art, of comic book making is hypnotizing
   others and drawing them into a [fantasized] real world ...

All the facts, descriptions, and metaphors of the session are now held
in this contextual scene as substantiation of the world of comic books. In
comics, there are more choices for Andy. Anything can happen and not
happen, without him having to get into trouble with the outside authorities. His creativity and imagination can flourish more but be held safe
within the protective confines of art.

M: A real ...

BK: ... that's not real. And so it stopped. So he probably tried to make his
   life a comic book. (Everyone laughs.)

This is the transformative moment. All the circumstances of his life have brought him to this fork in the road. He can continue being a tragic comic or initiate becoming a successful creator, an artist who creates comic books. All forthcoming conversation will be brought forth by this choice: to be an artist or not to be an artist, to become a creator of comic books or to be a comic book.

T: … will do it. Yeah.

BK: He's been acting like a comic book character. I don't know the details of your life, but I'm going to guess that you have been acting like a comic book character carrying on and doing things that a comic book character would do. Is that true?

S: Hmmm. Yeah, it gives me some ideas.

Act V: Freeing Your Hand

BK: Yeah. So you just need to pick up … Which hand do you use to draw? Are you right-handed?

M: Yeah, he's right-handed. I'm left-handed.

BK: Your mind has been waiting for your hand to start … When you make a comic book, do you use … pencils? What do you use? Is it a pencil?

S: A pencil.

BK: Do you have a special pencil to make a comic book?

S: Mostly any pencil I can write with.

BK: (To M and G) I'm going to ask the two of you perhaps the most important thing anyone will ask you to do with this young man. And that is: immediately after this session, go get him a special set of pencils and paper, so he can get on with his life and become a great creator of comic books.

This is a direct invitation to transform their lives. I have underscored it as the most important request they will ever hear. The whole family is invited to recontextualize Andy's life as that of a creative comic book artist.

M: Hmmm. Okay.

BK: Will you do that?

M: Yeah. Uhmmm.

BK: Because this is a hand that's waiting for a talent to express. When he was born …

G: I did that three or four years ago, but I guess he wasn't ready.

BK: Really?

G: Yeah.

BK: Maybe not, but you see …

M: See now he's ready.

BK: ... our creative minds do all kinds of creative things. When he has a hand that's wanting to express this beautiful imagination, this hypnotic way of influencing people, and it's not used, if it doesn't have the right pencil, what happens is he's going to turn into the very thing his hand wants to express. I guess the decision for this family is whether he'll be a comic book or whether he'll create comic books. I advise you to make, create comic books because that will give you a rich, interesting, maybe even lucrative life. The other [choice] will just make your hand want to have a pencil more than it already does.

Again, the transformation is set in front of them. In this session, Andy has become an artist. The family is being asked to continue living inside this resourceful contextual home. Here, Andy creates comic books. Otherwise, he becomes a comic book. Both are potentially hypnotic, wild, influential, startling, disturbing, and beyond imagination. An artist gets paid for being wild and crazy, whereas the other gets you into boring kinds of trouble. His situation has now become: if his hand expresses his talent, he will become an artist; if his hand is held back, he may "turn into the very thing his hand wants to express." Notice that I recast the latter choice as a disguised alternative: "The [other] choice will just make your hand want to have a pencil more than it already does." In this latter resourceful framing, he can choose to be an artist or choose to desire being an artist more than he does now, which paradoxically moves him closer to becoming an artist.

M: Make him do it hisself. (Laughs)

BK: See, in a comic book, you can say anything. You can do anything.

M: Yeah, really.

BK: You can set fires. You can go to the backyard and look at a dog and just take your pencil and say, "I change you into metal." You can see a deer walk by and take your pencil and go, "I change you into a five-headed octopus with a sparrow's wing and the sound of a bumblebee when you fly near people."

M: Don't tell him. He might do it! (Laughs)

BK: Because a pencil can do it. His hand can do it.

Either his hand expresses a "not really" or he enacts it. "Not really" has become the same as a comic book.

G: (To M) His hand can draw that.

M: I know. I'm ah, I'm telling him not to give him ideas.

G: Well yeah.

M: Not doing that to the dog.

G: Put some ideas in his head.

BK: But you know, you know.

M: No, no. (Laughs)

G: But he can draw it on a piece of paper. That's what he's talking about.

M: Oh.

BK: Is that weird? A creative man like this, [with creativity] in his hand, needs to draw these things. His imagination needs to be drawn. When that doesn't happen, then you do all kinds of crazy things, like try to really make a dog …

G and M: Oh yeah.

G: He's always had an avid imagination. I mean even when he was little.

BK: I'm going to give you a professional … I'm going to name what's happened. This is professional talk: His hand is constipated. His drawing hand is constipated. It has to be set free. The creativity that is in him has to come out. He has to draw. He has to make comic books. You can't tell him, "Don't do comic books." In fact, you can't tell him …

I give them an absurd diagnosis, calling it professional talk: "His hand is constipated." The session has now focused to being about releasing his hand to draw comics rather than being a comic.

G: No, I haven't.

BK: Don't tell him not to make comic books that aren't too scary. He should make the scariest comic books he wants to do. He should not only have them be scary, but they should be a little crazy … he can have a story where the thing that scares people the most ends up being the nicest person.

M: Yeah and something about knives and he could do something about knives …

The scary thing he previously did to his mother was claim he would kill her with a knife. Now she, while fully inside this alternative context, encourages him to make a comic book about knives.

BK: I'm talking about the comics. I'm just talking about the comics.

G: Yeah.

BK: That's right. I think comic books [often] get boring because they show the scary people to be the bad guys. The scary people should sometimes be shown to be those who are the smartest, [those] who see the world and know that they have to appear not as they really are in order to help move life in a way nobody expected. That was a … mouthful. But I think you understand what I'm saying, don't you?

I'm suggesting that there are many other ways of working with monsters, including the surprise that they may be the real good guys.

S: Uhmm.

BK: Excellent. I knew you did. I knew that you were born with a special mission, with a special talent. And your mission is to use that special talent. I'm gonna just take a brief recess and get some water for myself. I'd like for you [the therapist] to talk to them and for him to share what kind of pencils he wants ... what kind of paper? Maybe they should have a certain space in his room just for the ... Maybe they should put a sign over his bedroom door saying "professional cartoonist."

The family and therapist are left to more deeply settle into their new contextual home. They are to discuss the tools Andy needs for building a new life and discuss rearranging his room.

T: Cool!

BK: This all needs to come in their life in a big way.

T: Yeah!

BK: You all talk about it. I'll be right back. Okay?

T: Cool!

BK: It's awesome, man. (Patting Andy on the back, BK exits.)

T: Awesome! So, what do you think about that?

G: (Laughs) Amazing to me!

M: Amazing to me, too.

S: I should have brought ... I got my binder in my book bag. I can show you the rest of my pictures.

The therapist and family discuss the kind of art materials Andy wants. He specifies big pieces of paper on a pad and notes that his studio should be located in his bedroom, near his desk and window. Grandma recommends a setup where he "can look out the window and think and concentrate." They put together a plan to go both to a hobby center and Walmart.

When I come back to my office, I am thinking of a heavy metal arm that I have on one of my shelves. I frequent antique stores and flea markets to find unusual items that I can give to clients when the moment seems right. My office is filled with these unusual items. I enter with the decision to give him the metal hand.

BK: (Reaches on shelf for the metal hand) For years I've had a hand, not understanding why I had a hand. I know that you had noticed that it's a right hand. I think maybe this hand's been waiting for you and that I should give you this hand. Now when I give you this hand, your hand is going to be free to express itself fully through the art that's your talent.

I prepare to give Andy the free metal hand in a ritualistic way, concretely marking the moment when his hand is declared free to express his inborn gift and destined mission.

S: Yeah.

BK: Okay? (BK offers Andy the hand.)

S: Thank you.

BK: You bet.

T: Is it pretty heavy? (M is quietly chatting with G.)

S: Yeah (smile on face).

BK: It's heavy.

T: It's awesome.

S: It's steel, ain't it?

BK: Is it?

S: Yeah. It's steel.

G: Do you know what I was sitting here thinking about? The year before last, we went to church one night to a special meeting. The man that did this talking that night was a … preacher, but he'd had a stroke and a lot of things that he says is not clear. But he got up there and … he had comic books drawn …

BK: Really?

G: … that just hit it right on the nail. … I don't know who made the comic books, where he got it from or anything about it … Was sort of like these people dressed up and they would go out to some part of the outside in the woods in the, like in the country, like in the woods like—it was like ceremony that they would do, but then it started getting into the school, and they had to put a stop to it.

BK: Do you know what's interesting to me about what you said? (G and BK lean toward each other.) You found yourself listening to a preacher who for some reason brought up comic books. That's almost as if it's, I don't want to be too ambitious with my thinking …

G: Yeah.

BK: But it's almost as if already you were given a sign that comic books have something to do with the path of his future.

I utilize what Andy's grandmother presents as further validation of their new contextual home.

G: Well you know.

BK: (Looking at therapist) That's extraordinary, you know?

G: It was really amazing to me because I took the comic book home and I read it.

BK: Yes.

G: And then it came up into the schools. And it was happening at school, what they were doing and what they were saying and all this other ...

BK: (Looking at Andy) Now this is important. I want to say this to your talent. I want to speak to your talent: When you make a comic book, that becomes your whole world. Every comic book you make can be anything that your talent wants to express. That means you can create many worlds. (Andy nods in agreement.) You can sometimes base your world on things that happen in your family. I mean, you might just turn to your family and see it as an inspiration for the kind of world you can create in that comic book world. So, if mom almost burns the house down, it gives you an idea for a comic.

By speaking to his "talent," I am able to underscore, in a different way, the presence of his gift. Addressing the talent and calling it out by name validates it. I also suggest, consistent with what we have established earlier in the session, that his family is a source of inspiration (even when it's an annoyance). Whatever happens at home can be utilized to bring forth new creative expression in a comic book.

BK: If Grandma talks about a preacher who talks about a comic book, it might give you an idea. In a way, everything that happens in your life ... (G interrupts.)

G: Hey, I have another idea!

BK: Yes?

G: When he was small, we have a pond out back ...

T: Behind your house, yeah.

G: And when Pa Pa passed away ... he would take knives out and bury them and we didn't know where he buried them. We don't know what he did with them. But now it's all cleared-off land now. So, they may be there, you know, and they may not be. But, anyway, he would ah ...

BK: Who's he? Who buried them? (Andy puts up hand.)

G: (Pointing to Andy) He did.

BK: Okay.

G: And he would walk out around in the woods at that time and he would go around this pond and stare around. He would go further if I didn't go follow him. Maybe he could draw something like that.

BK: He can use anything ...

G: A little boy going out in the woods ...

BK: That's a wonderful idea. Anything that inspires him.

T: Yeah, Yeah.

BK: You know and sometimes you think thoughts that don't make any
      sense. You know what happens if you're ...

G: Well, ah, that ...

BK: One sec. Just let me finish this. Because I want to fully address what
      you just offered. A creative human being thinks things other
      people don't think.

G: Mm hm.

BK: What's important is they take that creative inspiration and express it
      in the art that God gave them to use. If he doesn't use his right
      hand, the hand that's been stuck—but now he is—what hap-
      pens? It just comes out in weird ways. Now we want those things
      to come out not only in weird ways, but even weirder ways than
      anybody ever imagined. Because he's free to do it in his own
      world, which is the world of the comic book. (Looking at Andy)
      Do you understand what I'm saying? (Andy nods.)

G: Well I think I've told you that about him going out in the woods. (BK
      nods.) I think he was looking for someone (G mouths "Pa Pa"
      without saying it out loud.)

Grandma is telling us that his grandfather's death was an important
marker in Andy's life. I would not be surprised if it marked the onset of the
acting out behavior that caught the attention of others. It is also interesting
to speculate, after the fact, over the connection between a preacher's comic
books that describe weird rituals in the woods and Andy's walking around
the pond and burying knives after his grandfather's death. As fascinating
as this is, a therapist must avoid being carelessly tempted to explore it. This
would most likely unravel the progression that has been made and lead to
a return to a "problem"-oriented theme. In this session, I look for how to
utilize Pa Pa's presence in the new contextual home.

BK: When you have somebody like that, that you love so deeply, you never
      get over it. You never get over ...

G: I think that's what, I think that's what he was doing.

BK: You never stop missing them. You sometimes have dreams where
      you even cry because you see them, because you feel them
      close to you in your dreams. Do you all ever have a dream
      like that?

M: I had one ...

BK: (Looking directly at Andy) Do you ever have a dream like that where
      you ... (Andy starts weeping.)

M: I have a dream ...

BK: ... where you see them and feel them and then you just start crying
      because you feel alone.

M: I had a dream ...

BK: Just a second (to M). Because I understand that because I had a grandfather like that. You know what? I still have dreams like that of both my grandpa and of my grandma. When I see them, sometimes I'll just see them sitting in a chair ... it's very interesting because once I had a dream where he actually buried a key in the ground and I dug it up and it was the key to his house where I used to go see him. ... It just means that somebody loved us deeply. We know that it's a special love because it will touch us that way when we miss them and they'll come, bring back the memory of their love in the dreams. I ... know that there was a great love between you ...

Andy's experiences that are connected to his grandfather—dreaming him with tears and burying objects in the ground—are associated with my own experiences. They are regarded as confirmations of the love we have for our grandfathers.

G and M: Mm hm.

BK: ... and that when he uses his hands to express the gift that God gave him to use, Grandpa would be very happy. That Grandpa would be, dare I say, smiling and dancing now. Do you know what I mean? (G nods.) You know what? My guess is that he just didn't come into the world with this creative gift from nowhere. This family is a creative family (quiet laughter). That is, on the one hand ...

I have brought grandfather into the newly built contextual home, where he can be imagined to be happy about Andy following his destiny to express his creative gift.

T: He's never bored. (Laughter)

BK: ... on the one hand, there is one thing after another, but not real.

G: (Repeating) Never what?

T: Never bored (smiling while looking at G).

G: Oh ...

BK: I bet there's some interesting stories about Pa Pa's life that would make an interesting cartoon.

G: I wished I could remember what he [Pa Pa] always talked about, joked about, ah, who were they (looking at M)?

M: I don't know.

G: You don't. Oh my mind just went blank. "Little Moron!" He used to talk about Little Moron. He joked about Little Moron.

BK: Interesting.

G: He would tell that and he could get into a bunch of people and he would be the center of attention.

BK: Isn't that weird? Because the first part of the word "moron" is "more," which is the last part of the name of the monster Andy wanted to create, which was Magmore.

This is another way of bringing Pa Pa into the home of an aspiring young artist. What is implied, but not made explicit, is that Andy's love for his grandfather provides a bridge, a connection, enabling an influence upon what he creates. The very name of Magmore holds a part of Pa Pa's imaginary joke character's name, Little Moron. We now have the possibility of "more" meaning for Magmore.

T: Magmore.
BK: That's really interesting, isn't it?
G: Oh (looking at BK), how did you …?

(Laughter)

BK: Isn't that a weird thought?
G: Yes!
BK: Whoa. That's how creative he is. He doesn't even know how creative he is!
T: Wow (looking at Andy)!
BK: (Motioning toward Andy while looking at G and M) He made a connection with his grandfather even in the name of what he was going to create as the character for the comic.
G: Oh, okay.
T: Wow!
BK: Wow.
G: (Looking at Andy) More, what did you say?
S and BK: Magmore.
G: Magmore. Moron. Magmore. Okay!
BK: So you got the "more." Half the word is there.
M: Yeah it is.
BK: Mm hm.
G: I wished I could remember some of those things that he used to tell.
BK: I think (pause) when he receives—because he's already received the blessing that his hand is now free to express the gift he has—when he receives the pencils and the paper, sets the room up, and steps into his new life of becoming someone who's here to express his talent, I think you're going to be surprised by the things you remember about Pa Pa (looking at G, then turns to M). And I think you're going to be surprised by the way in which there's a lot more creativity in all of you that will start to surprise you as you see his creativity flow. And boy when it comes on … get ready, 'cause … you better have enough paper. 'Cause

he just might be going to town. You know, it's up to him how much he wants it to flow.

Everything that is presented in the session is utilized to further elaborate and substantiate the reality of a family that nurtures an artist who creates comic books.

G: How did somebody like you two people come into our lives? I mean, I don't ... (Laughs)

BK: I was going to ask you ...

G: It's fascinating.

BK: ... how can fascinating, creative people like you come into our lives? Because it's such an inspiration ...

G: I mean ...

BK: ... to see how. ... Are you all religious?

G: Yes.

M: Yes.

BK: And would you say ...

G: No. He's not.

M: He's not.

BK: But you ...

G: I don't know why he's not, but ...

BK: (Looking at Andy) But you ... recognize that there's mysteries in the world, yes? We can say that there's mystery.

S: (Looks up and nods) Yeah.

BK: It's a mystery that all of these fascinating things take place ... (T nods.)

G: He doesn't ...

BK: ... in ways ...

G: ... he doesn't prefer to go to church. Why? I don't know.

BK: Well ... in his comic book, he is free to invent a church nobody's ever heard of. Wouldn't that be interesting? (Andy nods.) Yeah. The Church of Magmore. (Everyone laughs; Andy is rolling his head and laughing; BK then stands and shakes hands with M, then G, and then Andy.) Well, I think you all are going to have one of the most fascinating stories to share with the world. Maybe someday it will be a Hollywood movie. No, some day it will be a cartoon [comic book] in everyone's home.

One might assume that one of the biggest disagreements in the household concerns God and going to church. Andy's delight and explosion of laughter about the thought of creating a Church of Magmore shows how free he really can be when he creates a comic book world.

T: Yeah!

BK: That would be something, wouldn't it? You never know. Cartoons become movies, movies become cartoons, and lives become cartoons, and lives become movies, and movies become lives. What a movie you've been living. Now it's time to make it free to express everything that touches you most deeply, including the feelings of Pa Pa … you know, Pa Pa's in the family …

Their life has been unreal ("not really") like a comic book, a cartoon, or a movie. In these fantasy worlds, anything is possible, including the choice to be free to express feelings for the loved ones you miss the most. To an imaginative boy, the death of a close grandfather could easily feel as if life murdered a dear relation. Burying knives underneath the ground is a creative way of burying the instruments of death. Walking around the perimeter of a pond is as imaginative, and perhaps mythological. The circle of life—the water that holds and births and nourishes life—is surrounded by the possibility of death, which lies buried underneath the ground, waiting to bring others down under with it. As interesting as this fantasizing may be, I did not bring it into the conversation because what is most important is that we bring forth ways in which the presence of Pa Pa's absence can be a resource for the young aspiring artist.

M: Yeah.
BK: … 'cause you still feel your love for him.
M: Yeah.
BK: He still shows up in dreams. Pa Pa's still there.
M: I had a cousin got killed in a car wreck …
G: There's a hummingbird comes by my house.
BK: Really?
G: Every once in a while.
BK: (Looking at Andy) There you go. … Magmore's hummingbird … What could you do to a hummingbird in a comic book world that would just make people think real but, not real … that would be so interesting and hypnotic? Something for you to think about, you know? Maybe that would be a comic book … maybe for Christmas you'd make a comic book for each person in the family. Grandmother gets a hummingbird …

Again, I'm utilizing everything for the artist's imagination and creative expression.

M: And I'd get one with the house on fire.
S: And another thing …
BK: What was …
M: I'd get one with someone setting the house on fire.
BK: Yeah, she gets one …

S: I'm thinking about, I'm thinking, like, with the drawings and stuff, making a video game, too.

BK: Wow! Awesome. There's so many, you know …

T: So many ideas.

BK: When you take all the energy that you've been wasting on nonsense and creatively apply it to the world, it can make a surprise you're not going to believe.

The comic book–like behavior that originally got him in trouble is now regarded as "nonsense," a waste of creativity and energy that brings less surprise than what would be produced by being an artist.

M: It'll probably …

BK: And your life is going to be so (leans toward Andy) weird. (Andy looks up and smiles.)

M: It'll surprise everybody.

BK: And surprise everyone.

T: Yes, that's true.

BK: Your ability to hypnotize everyone will multiply 10 times. And your ability to disturb people in ways that make you smile will multiply 10 times. But you know what? You'll get away with it; and secondly, some day you might make a fortune doing it. But one thing's for sure: this whole family is already very rich, rich with the unique creativity that you've all brought together (M laughs), and come to us and shared, and that's just a blessing for us. I think you need, as quickly as you can, without breaking the speed limit, to go start the next chapter of this gifted young man's life. Okay? Turn his room into the studio of a professional cartoonist.

M: I didn't think about that until you said something.

BK: Okay? So you all should get on with it. Now you have three hands and there's the beginning part [the steel hand] for Magmore.

T: Magmore.

M: I never heard of …

BK: Maybe Magmore will be a monster who secretly, when nobody's looking, draws cartoons with a steel hand.

T: Whoa.

M: I never go in his room when he's drawing.

T: I'm eager to see what Magmore looks like.

M: I do, too.

BK: Awesome. Okay, that's it! That's it.

T: Great!

BK: Go make some history.

T: Awesome.

(Everyone stands.)

BK: Let me shake your hands. Really nice to meet you. All right, let me shake that metal hand, too.

T: (Laughing) Very good. Very nice.

BK: Great! (Laughs) Fabulous.

Andy: (To T) There's your ball.

T: Oh, thank you.

BK: Yeah, he doesn't need it; he's got a steel hand.

T: He doesn't need this. (BK laughs.) It looks hard but it's soft, really. (T opens door.) All right, there you go.

BK: Thank you, all. Take care.

We followed up for several months and found that Andy was successfully doing his schoolwork and the family reported that things were moving along nicely at home. His bedroom, now a studio, is a place where his imagination is able to express itself with a hand free to create imaginary worlds. Following a conversation with the therapist about historic knights, the family was invited back to the clinic to conduct a knighting ceremony. (How interesting it is that knights covered themselves in metal armor and that Andy now has a metal hand? He also has a pencil as a sword that can create—and end—any battle his imagination desires.)

During the follow-up session, I asked everyone to leave the room except Andy. I told him that I had done some research and discovered the official pledge that a squire must take before becoming a knight. A few things had been added to the pledge so it would be unique for his life. I asked if he and Magmore would like to be co-knighted. He affirmed that he would, and I asked him to stand. As I proceeded to turn him counterclockwise, around and around, I announced: "People, especially adults, say all kinds of things that don't fit you, and you don't have to hold onto that. As I spin you, let's unwind all of that. Let it go. Let's rewind and erase it."

I then stopped turning him and went on to say, "Step into your new life with a new beginning. Now you shall become a knight. Lie down on the floor."

Without hesitation Andy went down to the floor, where I covered him with pillows. I walked to the door where the room light switch was located and asked both Andy and Magmore to verbally indicate their acceptance of each part of the pledge. After each promise was read and agreed to, I turned off the lights in the room, allowed a moment of darkness, and then turned the light back on. This use of lighting continued throughout the entire pledge:

"We pledge to make sure Magmore always defends a lady."
"Yes."

(darkness)

(light back on)

"We pledge to make sure Magmore only speaks the truth, even if others don't understand that it is the truth."
"Yes."

(darkness)
(light back on)

"We pledge to make sure Magmore is loyal to his creator, Andy Jackson."
"Yes."

(darkness)
(light back on)

"We pledge to make sure Magmore is devoted to all his missions."
"Yes."

(darkness)
(light back on)

"We pledge to make sure Magmore will defend the poor and helpless."
"Yes."

(darkness)
(light back on)

"We pledge to make sure that Magmore is brave."
"Yes."

(darkness)
(light back on)

"We pledge to make sure that when Magmore is on a quest, he will remove his armor only when he is sleeping."
"Yes."

(darkness)
(light back on)

"We pledge to make sure Magmore never avoids dangerous paths out of fear."
"Yes."

(darkness)
(light back on)

"We pledge to make sure Magmore is on time for any engagements of battle."

"Yes."

(darkness)
(light back on)

"We pledge to make sure that, upon returning home from any adventure, Magmore will always tell someone of his escapades and adventures."
"Yes."

(darkness)
(light back on)

"We pledge to make sure Magmore only fights one-on-one with an opponent."
"Yes."

(darkness)
(light back on)

"We pledge to make sure Magmore is funny from time to time."
"Yes."

(darkness)
(light back on)

"We pledge to make sure that Magmore will tease other people."
"Yes."

(darkness)
(light back on)

"We pledge to make sure Magmore is wise and makes it impossible for others to trick him into getting into trouble."
"Yes."

(darkness)
(light back on)

"We pledge to make sure Magmore invites other comic book characters into the life of Andy Jackson."
"Yes." (with laughter)
I respond, "I like the way you said 'yes' to that one. Are you aware that that was your most powerful 'yes'?"

(darkness)
(light back on)

"We pledge to make sure Magmore sees to it that Andy understands that his pencil is his mighty sword."
"Yes."

(darkness)
(light back on)

"We pledge to make sure that Magmore sees to it that Andy's mighty pencil-sword is always sharpened."
"Yes."

(darkness)
(light back on)

"We pledge to make sure Magmore sees to it that Andy's comic books are as powerful a battle as any fought with blood and guts."
"Yes," Andy says with an unusually loud response.
I address this, "I like the way you said that strongly."

(darkness)
(light back on)

"We pledge to make sure that Magmore and Andy always remain loyal friends to one another."
"Yes."

(darkness)
(light back on)

"We pledge to make sure that Magmore and Andy will stay close even if Magmore finds and marries a Mrs. Magmore and becomes a father to little Magmores."
"Yes," he replies with a chuckle.
I add, "That's sort of a crazy one, isn't it?"

(darkness)
(light back on)

"We pledge to make sure that Magmore will be alive even after he passes on to another world."
"Yes."

(darkness)
(light back on)

"We pledge to make sure that Magmore and Andy will know each other even when Andy is an old man who can barely hold his pencil sword."
"Yes."

(darkness)
(light back on)

"We pledge to make sure that Magmore always will be the protector of Andy and Andy will always be the protector of Magmore."
"Yes."

(darkness)
(light back on)

For the next pledge, I prepare Andy, "Are you ready for this one because it is strange?"
"We pledge to make sure that, in some weird way, Magmore and Andy always will be Louisiana homeboys."
"Yes," he responds while giggling with delight.
"Excellent! I warned you about that one." (We both laugh.)

(darkness)
(light back on)

"We pledge to make sure that you always remember this day."
"Yes."

(darkness)
(light back on)

I ceremonially removed the pillows covering Andy, one by one, and lifted him up, while officially declaring:
"Let it be known that these things are true inside and outside of all the comic books that will be created by these two knights. Let their mission be to spread joyful surprise and unexpected merriment to all those who honor the mysteries of the comic lords. I declare this by the power invested in us by the release of the sacred metal hand. And so it has been declared."

"Congratulations. You have been knighted." After I shake his hand, we hug, while I continue, "You've been knighted in a way that only you and I can understand. Now you know what to do with the rest of your life. ... Maybe someday you'll put this knighting ceremony into one of your comic books."

We then invited his mother, grandmother, therapist, probation officer, and supervisor into the office to have a party complete with soft drinks and chocolate cake. He was given a new fishing pole that his mother used to officially knight him as he kneeled. Now he can sit on the banks of their pond and fish for new ways to inscribe an honorable, creative life. Sir Andy and Sir Magmore are now creating worlds never ever imagined before they came into our world.

Several months later, Sir Andy initiated getting a job on his own and volunteered for public service to clean graffiti off of public walls.

He's still drawing and doing well in school. As a final follow-up, my colleague and I went to his home during the Christmas holiday and he was proud to show us his grades. He had made all A's. We celebrated how his story can provide inspiration to other adolescents whose natural gifts are not always seen or understood by others. He invited us to see his studio. We took a glance at the spot where he created his cartoon figures. There he gave me his first drawing of Magmore as a gift:

Figure 12.1 Magmore. Reprinted with permission.

To my surprise, his drawing was the image of a Balinese magical figure. Around five years prior, I had written an obscure book with I Wayan Budi Asa Mekel on the magical drawings of the balians (healers) of Bali (Keeney & Mekel, 2004). Andy, without knowing anything about this culture, had drawn something very similar to one of their powerful images, that of Sang Hyang Bhima Sasa-Dhara. There, a traditional healer draws this image on a lontal (palm) leaf, places it in a glass of clean water, surrounds the client with fragrant flowers, and then asks him to drink the water and clean his face with it. Its purpose? The drawing is made and utilized in this traditional way in order to transform anything inside a person that is toxic so he will no longer be sick.

Magmore was more than a comic book figure. It was a reminder that therapy will always be a mystery.

CHAPTER 13

Christine Apple Nut's Theatre of Therapy

Creative therapists are most creative when they draw upon their own natural resources, skills, gifts, talents, and eccentricities. If a therapist is naturally funny, then humor should be accessed in sessions. If one has the skill of inducing boredom, then draw upon it as a means of trance work. The next chapters present cases where I have worked with therapists in order to help them access their own unique resources as therapists. I call this work a "therapy of therapy."

The life of being a creative therapist is inseparable from the quest to becoming a creative human being. Awakening our clinical work requires revitalizing our presence in everyday life. The following story shows how a unique creative style of therapy was brought forth in an unexpected way. Here, a high school counselor literally ends up establishing a theatre in her own counseling office at the school where she works.

On November 29, 2007, I was introduced to Christine Abrahams, a school counselor in New Jersey. She wanted to do some transformative work with me via e-mail correspondence. In an initial phone interview, she said that she wanted to shake up her routines. She had an office at school that was filled with wild things, like a witch's head in a globe that talks, a neon clock, and a small mannequin in a pink dress that had been given to her by her colleague, Richard, whom she described as a "court jester."

She described herself as a "nut" and said that her boss called her that. She secretly wants to be a novelist and has written four drafts of a first novel. In her own words, she "wants to see past the doors" that lead to greater imagination. She explained that her favorite color was "blue" because it connected her to the sky and ocean and she enjoyed the sense of expansiveness that color brought her.

Act I: Remember to Clean the Battery Contacts

I asked Christine to think about what she had recently done that was spontaneously playful and let me know by e-mail. The next day I received the following e-mail letter from Christine.

November 29, 2007

Brad,

I had trouble answering the question about what I did today that was—I think you said playful or spontaneous? Well, I was speaking to my friend about my day and realized that I did something pretty trickster-like but hadn't remembered because I think I was taking our conversation too seriously and since I do this stuff all of the time, it didn't stand out.

A colleague of mine, who I'm [clinically] supervising, asked me to fix some computer gadget he had. I suggested that he clean the battery contacts, a task my husband tells me to do all the time with non-working gadgets. I added that my husband would be proud that I remembered something he told me and that if he died tomorrow, I'd probably be able to build an entire house because I've watched him doing it over the years.

Well the intern-counselor asked who I thought would die first— me or Bruce, my husband. This guy is such a Boy Scout that it's hard to get a rise out of him, so I said that it would depend on who succeeded first in doing the other one in. His eyebrows rose. I explained that I continually try to do him in by strategically placing cat hairs around the house to inflict an allergic reaction, while my husband constantly tries to poison me by putting unknown and potentially deadly food substances in the food he cooks me. I told the intern that I'm immune because of all the Qi Gong that I practice.

He stared at me for 30 seconds, got up and said, "Well then. Have a great evening," and left. I got a big chuckle out of that one. To top it all off, he's getting married in February.

Hope this helps!

Christine

I replied as follows:

November 30, 2007

Hi Christine,

I am happy to know more about your "trickster" side. For your first exploration in evolving your creative edge, I suggest that you cultivate this particular fantasy: your present job is providing you with interesting characters, scenes, and experimental interactions that will later contribute to your writing of nonfiction. Pretend that this is the case without any logical reflection or evaluation.

Your office is the scene of the main character's daily work. It is already off to a good start, but we are going to take things further. Begin by making a sky blue sign that says:

REMEMBER TO CLEAN THE BATTERY CONTACTS

You can make this sign in any way in any medium from paper to wood or metal. Think of starting a collection of battery contacts that can be on or near this sign. Place all of this in your office and make the words your new professional mantra. Bring in a small collection of batteries. Look for an old doorknob that can reside within the batteries.

Most importantly, have two piles of nuts (both a mini-bowl of edible nuts and a mini-bowl of metal nuts) inside this collage or assemblage or arrangement. When people ask you about this collection, hand them a nut and say, "This is an art assignment that is set up so that when people ask about it, I am able to say to them, 'Please consider being a nut.'" Tell them that is all you can say because it is a class assignment. Say no more.

It may be that the whole arrangement of sign, battery contacts, batteries, nuts, and other added items begins to look like a kind of trickster altar.

Whatever problem enters your office, whether carried by students, parents, or colleagues, make sure that you find a way to advise them with these words, "Remember to clean the battery contacts." You will have to consider all kinds of rationales and stories in order to have "battery," "contacts," and "cleaning" become transformative metaphors.

Consider that this activity constitutes planting the seeds for a garden of stories.

You may also consider having some small business cards made up with the same words: "Remember to clean the battery contacts." This way you can have more practice weaving the advice into conversations with people you meet on the go. Be prepared to hand them the card at the appropriate moment.

You are on your way to shaking up your world with words that are specifically unique, important, and rooted to you. You will be using all the professional parts of your creative self—brokering words that teach and counsel others with literary imagination and transformative possibility. Weave the stories and ideas that start to emerge into your own ongoing novel or start a collection of brief short stories.

See yourself as a scatterer of crazy wisdom apple seeds or apple nuts, spreading the "cleaning of life's battery contacts." Maybe you will become the next Johnny Appleseed. Did you know that he not only spread apple trees but that he was also a mystic? Consider writing your possible future name on the back of your new sign, knowing that only you know about it:

"Christine Apple Nut"

At least make the motions of writing about it with an invisible pen.

I think it is a good idea to ask Richard to be your creative consultant and possible co-conspirator in this experiment in the transformation of everyday moments that can take place in a double agent's office. Finally, the two of you need to name the mannequin and find a way to link it to the "cleaning of battery contacts in everyday life."

Best,

Brad

December 8, 2007

Dear Brad,

It's been an interesting week. As soon as I received your letter, I designed a poster and glued batteries to it. I put out my nuts/bolts and a bowl of cashews. The responses have been interesting.

The students, whom I'm meeting for the first time, would take secret glances at the poster while I was writing something or looking at my computer. They didn't say anything. At the end of my session, after I asked them whether they had superpowers, I directed their attention to the poster and asked them what they thought it meant for them personally. Most of them didn't know what battery contacts were, but once I explained it, a lightbulb went off and they would apply it to whatever needed work in their lives at the moment. Here are some of their responses. I asked those who couldn't think of anything to ponder the saying and e-mail me their ideas. I'm planning to type up each response, paste them to colorful paper, and then attach them all to each other to help my poster grow.

What the students said:

"If you have dirty battery contacts, it won't work. Like if you have a bad record, your life will not go as planned. Clean contacts = doing better."

"Clean the battery contacts and fix the problem. This means you try harder. The doorknob: If the door is locked, you'll be late. If you unlock the door, you get to school early!" (This kid is always late to school!)

"Let go of your problems (clean the contacts) or they will drag you down."

"Cleaning the battery contacts means starting a new day with a clean slate."

Now of course I have staff popping in and out of my room and they are not afraid of asking what anything is. My boss snorted and said "figures." The science supervisor walked in and immediately read the saying out loud and then began to free associate:

"The door (knob) to the other dimension where you meet the nuts and bolts of life and the nuts in general. Keep an open mind ... keep the contacts clean." Then he shouted with a grin, "You have an altar of the absurd!" I offered him a nut and gave him permission to be a nut, to which he replied he already was one.

My friend Joan darted in and out all day, not saying anything about it, pondering its meaning. She asked me and I said, "How the hell should I know?" She offered her interpretation: "The battery is the energy of life. Battery contacts are the people and things that you provide energy to and vice versa. Doorknob is the entrance into understanding people. Nuts and bolts are the basics or things that hold everything together. Nuts = nourishment or fun—being a nut type of thing."

My other friend Debbie came, narrowed her eyes, and said, "Why do you have religious symbolism in a public school?" I didn't answer, not knowing what to say. "This is ritualistic looking," she said suspiciously. "Perhaps it means to clean your brain out—the neurotransmitters and with a renewed interest in your world, you can let go of old baggage." Then she left. You need to know that she's the school's drug and alcohol counselor.

The whole thing was hysterical! Just hearing the interpretations of folks coming in and out was a hoot.

Richard thought the whole thing was hysterical and immediately bought me another mannequin to go with Priscilla, the name we gave to the pink one. The new one is named Violet.

I found that even my special education kids were intrigued by the poster and what it could mean as a metaphor for their lives. One boy commented that it was the most interesting thing he's had to think about in a long time.

Thank you!

Christine Apple Nut

Act II: Apple Nuts

December 9, 2007

Dear Christine Apple Nut,
It is very nice hearing how your office has become an even more fascinating creative fertile ground for people to enter, pause, and wonder. Your idea of growing the poster is excellent. I think you should definitely continue to gather and post responses as they come into the office. In doing so, make sure that the key word and phrases in each response are underscored so that they stand out in a prominent way. The key words or metaphors supplied inside each of the responses you just reported include:

Let go
Clean slate
Figures
Altar of the absurd
How the h*** should I know?
Energy of life
Doing better
Fix
Nourishment of fun
Clean your brain out

Call these metaphors "the seeds."
Now when you gather at least seven seeds, join or string them together so they articulate a new insight for the world. For the above collection of seeds, we could creatively write something like:

"Let go of your clean slate with figures that honor the altar of the absurd. How the h** should I know? Because the energy of life said that doing better requires fixing some nourishment of fun so that you clean your brain out."

Now you can tell people that each person's response contributes to the making of a whole insight. Call each whole insight an "apple nut" and then say that you are growing "apple nuts." If anyone asks what an "apple nut" means, immediately ask them what they think it means. You know what to do with their responses: when you get at least seven of those seeds, make another string of connections that yields another apple nut.

As the garden in your office blossoms with more responses, more insights and wisdom will gather. Always remember to gather at least seven seeds before making a new apple nut.

Finally, make sure that the apple nuts (the strung-together seeds that create a whole crazy wisdom insight) are written on a piece of red paper that is shaped like an apple. Hang the apple somewhere in your special space.

Imagine the apples and the poster and all the rest becoming an apple tree, that is, an apple nut tree, in your office.

Make sure you keep a fresh supply of nuts and add an apple from time to time.

One more thing I almost forgot: start thinking of ways Priscilla and Violet are involved in this office adventure. Perhaps they are the guardians of the apple nut tree, or the gardeners, or the angels, or the trickster queens. Think of a way that they can communicate these words: "I'm no dummy."

On with your mission, with the deeply held (but never examined) knowing that this is important creative activity of the most transformative kind!

Best,

Brad

P.S. Please tell Richard to keep up the good work!

December 15, 2007

Dear Brad,

I'm continuing to collect the wisdom seeds from the students and will create apple nuts from them. This week another of the drug and alcohol abuse counselors created a very unique wisdom that she pasted on multicolored paper. The best part was that she signed it "SAC – Battery Recharger." She's getting into the swing of it.

I've been trying to incorporate the phrase into my discussions with my students. One student came to me complaining about her science teacher. While we were discussing her issues with him, I said, "Hmmm … Mr. Wilson sounds just like your father." She stopped and said, "Oh my god, you're right. Do you think that's why I can't stand him?" I said, "Maybe. So whenever you start to feel that intense dislike toward him, remember to clean the battery contacts!" She got it.

Priscilla and Violet are doing well and are guarding the wisdoms. I've also told the students that when I'm not in my office, they can consult with either P or V, as both ladies know everything anyway.

For sophomore year, I'm going to ask all of them again what super powers they have and then post those.

Have a wonderful weekend!!!

Christine Apple Nut

Act III: Getting Recharged

December 19, 2007

Dear Christine Apple Nut,

The time has come to move to the next level of creative activity within your office space. I would like for you to purchase a set of battery jumper cables. Find a way to connect one cable to Priscilla (you have to decide whether she is the positive or negative terminal) and another cable to Violet. The cables may be clipped on or attached in any way to the guardians.

As you create more and more apple nuts for your apple tree, know that your room will become fuller of unexpected wisdom, surprise, unknowing, and the air of transformation. Know that people will be coming to your office to clean their battery contacts whether they know that or not and whether they understand that or not.

When you find someone who truly gets what is going on, you'll be able to comment and ask them from time to time, "Are your battery contacts clean today?" Tell them that there are many ways to get a charge or recharge in life and that a person can only receive a charge when the battery contacts are clean. Otherwise the charge can't get through.

Here comes the interesting part: when you hear someone laughing (in a good way) at one of your apple nuts, then decide whether you think they should be let in on the next thing. If you feel their battery contacts are clean and shiny, then take them over to the guardians, grab the battery jumper cables, and while holding them, say "that giggle was a nice recharge." Tell them how you are learning more and more about how important laughter and silliness are to revitalizing a person's life. Not only does it help heal physical illness, it also helps awaken our innate abilities and good will.

Try this out with Richard and see if the two of you can make modifications to this protocol. Whatever makes the two of you giggle is what you should pay attention to doing with others, those who have clean and shiny battery contacts.

From the seeds, the apple nuts will come. So enters the apple tree. In this chapter of creative unfolding, a new way of making connection, contact, charging, recharging, and restarting will come to be.

Enjoy all the surprises that will come your way.

All the best,

Brad

January 19, 2008

Dear Brad,

Priscilla and Violet are in place with their jumper cable, guarding the apple nut tree. What's happening is that the students come in and see this crazy set up and laugh immediately, which gives me the opportunity to talk about getting "recharged." What I'm finding is that the students who have thought about the battery contacts phrase do come back to it when we're talking. I'm also finding that I'm able to introduce the phrase when students bring their "problems" to me and ask them to remember it when they are having the same difficulties.

Faculty is now getting in on it. Some think the whole thing is absolutely ridiculous; others like it and find it creative. But across the board, they don't see the relevance to their lives. And, sadly, some seem jealous. The kids are a lot more open to the process.

I'm going to continue to get the student's interpretations and from them create more apple nuts. The special ed kids really get the whole thing and they come up with the most interesting responses.

Looking forward to hearing from you.

Christine Apple Nut

Act IV: Theatre in the Office

January 24, 2008

Dear Christine Apple Nut,

I am happy to hear about the students who are delighted with what they see, hear, and experience in your office. They clearly are getting recharged. You are bringing good transformative process into their lives.

Now let's consider the rest of the world: those adults who have to decide whether they are jealous, confused, or delighted. Of course the kids are more open. They are the hope for the future. They need adults who have not lost what they still embody regarding the creative spark, sense of play, and sensitivity to what lies beyond rational holding.

Now what is important is that your office is becoming a lab for how to be a creative presence in the world. You are learning how to bring forth transformation in full view of those who get it and in front of those who don't get it. What you learn, both consciously and unconsciously, in your office applies to all other spaces in your life.

Let's add something to your office world. At the moment, everything is openly and publicly revealed. Let's start concealing some things. I would like to see you introduce something into your office that I concealed. I wish you could find one of those toy theatres they had in the Victorian days. You would have a stage sitting in your office and the curtain would be drawn. Maybe you can find one or make one. Or consider having a shoe box or other kind of box that has the lid removed. Set it up like a theatre stage and get some cloth to cover it. Have it sitting in your office with the curtain closed.

Now people will see something that has been added to your office but they will not know what it holds inside. Its performance is secretly held behind the curtain. The curtain will be closed except for those moments when you decide to open it. For those who enjoy and "get" the apple nuts, they may have the curtain opened. When you open it for them, have a card sitting on the mini-stage that says, "Thanks for being an apple nut." For those who don't get any of this, the curtain is not to be opened.

You can place other things on your stage or other sayings or nothing at all. This is the beginning of your having another reality within your present reality. And it is the time when you start selecting who gets into/onto the next stage of Christine's Apple Nut Theatre.

Enjoy!

Brad

February 27, 2008

Dear Brad,

A lot has happened. When I read your e-mail I thought how wonderful to have a theatre in my office, then groaned because I am not great at all with the arts and crafts part of these assignments. I had printed out your e-mail and it was sitting on my desk at home. I left to do something and was pondering how to create the theatre.

When I got home, my husband said, "So, how are you going to make the theatre?" See, he has no idea that I'm working with you because he's not too open to this stuff and truly thinks I'm nutty—so the less I tell him about all this the better. So, I didn't miss a beat and said, "You know me. I'm probably going to do my best with a shoe box." "I thought so," he said. Then he offered to make me a theatre. He's in commercial construction so this is right up his alley. We sat down and figured out the dimensions, and I chose the colors to paint it—purple on the outside and black on the inside with deep, red wine curtains.

He did a great job. When it came to the curtains, I felt that would be another challenge—I don't sew, or rather, I sew, but only to repair holes in clothing. Well, we were in a store looking for something curtain-like, when we both thought of the same idea to use fancy napkins! They worked perfectly and the only sewing I did was hemming them so they could be slipped onto the curtain rod my husband inserted into my theatre.

I placed what looks like a brass music stand (which is really to hold business cards) and chose the quote from Shakespeare: "There's nothing good or bad, but thinking makes it so." I put that on the music stand and placed a gnome next to it. That's it.

I brought it to work and showed Richard, who loved it. Many of the students saw it but wouldn't even ask. I think they are afraid of asking because there is so much weird stuff in my office. A friend of mine who is also a colleague walked into my office and asked, "What's that?" Before I could answer, she opened the curtains. I laughed because she was acting like a little kid. Another teacher asked to see what was behind the curtain and loved the quote. He now brings other teachers to visit my office to show them what I'm doing.

I revealed what was behind the curtain to a number of kids at what seemed to be the right times for them given what they had been telling me. Most of the kids I see are 14 to 15 years old. One young woman looked at the quote, and it made her smile and opened up a conversation about judging other people.

I think good work is going on.

Some disturbing stuff though: I might be asked to move into administration so I asked an administrator whether I would be able to bring my purple, confetti chairs to use in my office. She shook her head sadly and said no, probably not. This really sucks. I knew the answer before she said it. It seems that being in administration is designed to suck the joy and life out of you. It's so serious!! … I think the fear is that I won't be able to be silly and fun if I move on.

Well this apple nut is tired. The theatre is really cool.

Much love,

Christine Apple Nut

Act V: Invisible Theatre of Transformation

March 2, 2008

Dear Christine Apple Nut,
Thank you for your recent update. How thrilling that your husband has helped you create your very own theatre for your office. That is great news! I look forward to seeing it, curtain and all.

Please consider the possibility that your likely move to administration is being arranged by the gods. They want you to go undercover, becoming a kind of secret agent for their cosmic play. Since this may happen someday, now is the time to start getting ready for the future. Here is what I would like for you to do:

Continue learning from all the interesting stuff that lives in your office. Have some fantasies of bringing all of it into your next office, but this time it will have additional magical properties. Its weirdness, purple colors, and the like will only be seen by those who know how to see it. Everyone else will only see boring furniture. After you have exercised this fantasy for at least three times, proceed to the next steps:

Find or create a magic wand. Know that it can always be kept inside your desk drawer. When waved inside any room, you can close your eyes and see things in their full weirdness independent of how others can or cannot see it.

Next, start taking photos of everything inside your office, taken from many different perspectives. These photos will be placed inside an interesting scrapbook with the title, "The Other Office: Please Enter." Arrange all the photos in this book so that when someone looks through the pages, they feel they have entered another place. Now you can sit this book on your shelf and open it up whenever you need to have all your mojo around you. And remember, it can also be shared with others.

Now here is the one that will really rock the room: get copies of the photos of the purple chairs and anything else you think would never be allowed as furniture in an administrator's office. Cut out the images of the items and then place them in a setting somewhere in your office. Consider this your tiny office — an office of what is forbidden that has managed to get into your office by becoming flat and tiny.

Think about how many different ways you can bring the reality of your present office into future offices as well as other places. Maybe you should carry a mini-office in your purse, have one in your suitcase for traveling, place one in your refrigerator, or even have one hanging from your rear view mirror.

The office you have now built and are learning from never has to go away. It needs to start shape-shifting and morphing into other forms that can be placed into more places in the world. These morphs are different kinds of seeds, or should I say "apple nuts"?

Please enjoy yourself as you bring more theatre, surprise, weirdness, play, confusion, contradiction, ambiguity, fun, joy, and crazy wisdom love into all the worlds you can co-create!

Thank you for your serious and good nutty work,

Brad

April 4, 2008

Dear Brad,

The assignment is accomplished. … I have a feeling deep down that I really don't want to be an administrator. But, I did take photos, made the book, and wait for whatever comes next. However, my thinking and practice has changed regarding how I work as a counselor. A student came telling me that she's an uncontrollable gossip and I made a button for her and five of her friends. The button has a gnome with his finger to his lips with the words, "Gossip-Free Zone." I created a letter to the student and a story to go along with it. When I met with her afterwards, she said that although she's not gossip free, she's aware when she's doing it. I think that's a terrific first step.

Another young lady, whom I didn't know, came in and dumped this whole story on me about her terrible life. We brainstormed where else she could dump her junk and came up with an overnight bag that could have a label that says, "Shirley's Dumping Bag." I invited her to get such a bag and leave it in my office. She could then come in from time to time and dump whatever she likes in it. She seemed delighted that we could invent such an idea.

Another of my students, Diane, is miserable at home. Her grades weren't good, but she told me that she wants to be a nurse. We have a competitive program run with our local community college that guarantees a student a seat in a four-year nursing program if they get through our program. But, you must have a B average to get it. We talked about how this program would be her "golden ticket" out of her house and we decided to hunt down some gold foil and write, "This is my golden ticket out." She now looks at it when the going gets tough.

I used my wand today on a number of kids and I have one frequent flyer who pops in twice a month to grab the battery charger and repeat, "I am recharged" three times. His grades are improving.

I met a woman at a professional conference who has taken me on as a sort of apprentice to handle the overflow of her patients. She specializes in anger management and is teaching me what she knows. It's a bit too behavioral for me but our arrangement is great and she's terrific. I had my first client last night and was asked to bring a business card, which of course I didn't have, except for one that says, "Remember to clean the battery contacts." At first I thought that it was too silly to hand out to such a potentially angry client, but I had no time to make a new one. So I just added my MA, LPC, and NCC to it (I figured that this would give me some validity). At the end of the session, I handed her the card. She burst out laughing! I was relieved and thrilled. We agreed to discuss the meaning for her the following week. I did all the behavioral stuff that my mentor wanted me to do, but it didn't feel right, so once I dispensed with it all, it seemed like my intuition kicked in and we had a great session. She seemed to have more confidence, appeared less stressed, and had made some interesting connections by the end. And she left laughing.

I'm feeling more comfortable with my loony side and letting it come out more and more. Who cares what others think? I'm having fun and of course Richard is right there trying to out-loonitize me.

I'm thinking that the job I'm doing now just needs to be done as well as the private practice opportunity that simply fell into my lap.

I want to thank you for all the wisdom you've shared. It's been a wonderful ride.

I appreciate all the guidance you've given me and feel that I have not only benefited, but all those whom I came in contact with have benefited as well.

Much love to you,

Christine Apple Nut

Act VI: Apprentices

April 6, 2008

Dear Christine Apple Nut,
Any "deep down feeling" is very important. Since you are now awakening your creativity, it's no surprise that life is protecting you and leading you to where you need to be. Thank you for the reports about your important transformative interactions with students and clients. Most importantly, you are listening to your "deep down feelings" in sessions, paying attention when they shout that something is bullshit or not working. Listen and follow that director of your deepest interior theatre.

I am still very impressed with your custom-built stage. You are very fortunate to have a set designer in your household. Have you found a miniature doll that you can name Richard? This would enable you to put Richard on the stage. You would then have a co-therapist in your room at all times.

Now is the time for you to look for someone you can teach what you are learning—learning that you already knew but needed to have experiences to confirm its truth and authenticity. How can you prepare yourself for an apprentice? Should you begin with an imaginary apprentice? I suggest that you add another miniature doll or figure and set it next to the Richard doll. Be sure to name it and pretend it is your first apprentice.

How nice it would be to actually apprentice an imaginary figure. If you can apprentice this part of your imagination, you will discover that, in a way, you will be apprenticing yourself. You will teach yourself more things than you ever imagined.

Enter the next world within your worlds, where you find the teacher and the stage and the theatre, and the world that is deep, deep inside your feelings.

Much joy and love to the ones inside who are deeply familiar with the one called Christine Apple Nut,

Brad

June 7, 2008

Dear Brad,

I wanted to give you an update. I have a "Richard" apprentice. I bought a Harry Potter doll and pasted Richard's face on it. The students love it. I've also made my garden gnome my apprentice. Since I've created an apprentice, more of the counselors are seeking advice from me more and more. The students also seem to be more open to alternative ways of thinking about their lives and have made terrific progress.

The big news, however, is that I was just offered a job as Supervisor of Counseling for a neighboring district. During the interview process, I was really open with them about my out-of-the-box self, including magic wand and stage. They actually came to my school to interview my boss and colleagues about me, which is when I showed them my office. I felt that they needed to see "me" and because it didn't matter to me one way or the other about getting the job, I wasn't inclined to be somebody else. I guess they like "me." I have to work out the details with them, but I think I'm going to do it. I'm really, really nervous because it's a lot of responsibility and I have to work more days, but the challenge of it appeals to me and I have so many ideas about projects I'd like to do that it would be great to implement new things for the kids. I also plan to bring my purple confetti chairs and stage.

I'm excited about it. Thank you again for all of your help.

Much love,

Christine Apple Nut

June 8, 2008

Dear Christine Apple Nut,
Congratulations! This is very special news because they hired all of you, including the Apple Nut. That is making a very special kind of history. I am proud of you. It is sort of a perfect graduation for our work with each other.

The apprenticeship idea shows that you are well on your way to always having an interesting creative future.

For your last consultation, I would like to recommend that you gather a collection of dolls for your office, perhaps cutout paper dolls. Every time a student comes to you for counseling, you can first see how they fit into your special world of surprises. When the moment is right and when it feels appropriate, tell a student that you think that their "problem" or "issue" or "reason" for seeing you was not the real reason at all. Instead, they were somehow magically brought to your office to see whether they could become an Apple Nut Apprentice.

Then get out a new doll and write the student's name on it. Perhaps attach a face shot onto its head. You will know how to continue with this. It may be different for each person or some things may be the same while other details are varied. You will know because you know you are the one and only true Christine Apple Nut.

With this new reality, some of your students will qualify to become your apprentices. The same may happen to some of your counselors. Before you know it, your stage will be very busy and your tree will be very full.

You are becoming just like Johnny Appleseed. Christine Apple Nut has set up her beacon light, theatre, with all the stage props. The Apple Nuts are coming. The Apple Nuts are coming! Get ready to change people's lives in ways that you have only begun to see. Get ready to make history, I mean *herstory*, I mean *applenutstory!*

It has been a delight working with you.
Much love,

Brad

Christine Apple Nut's Theatre of Therapy can inspire other therapists to build their own unique theatres. As we've seen, such a theatre may physically sit in an office in full view of clients or be hidden in a scrapbook, miniaturized, folded inside a wallet, or invisibly tucked away in imagination. What is most important is to contextualize your clinical work as performance and place it on a dynamic, vibrant stage that draws upon creative expression while serving transformation. Christine Apple Nut now helps other counselors be more creative and theatrical in their school offices. She also published her first novel.

The professionalized use of the arts in therapy, sometimes called "expressive therapy," unfortunately turned things upside down. We don't want music and painting inside the frame of therapy so that we get "music therapy," "art therapy," "dance therapy," and so forth. Therapy should not be the context for aesthetic expression. We want therapeutic and transformative presence inside the frame of art. What we desperately need is transformative art.

These days, my work is involved with both clients in traditional mental health settings and with the "therapy of therapy" conducted with therapists and other transformative artists. The latter aims to help therapists secretly move out of therapy and move into a seen or unseen creative theatre of transformation. This doesn't mean that onlookers see less therapy. What they see is more therapeutic presence, spontaneity, improvisation, and creative play. What disappears includes stereotyped conduct, robotic clichés, habituated hypotheses, and mindless trivialization of clients through psychiatric labeling, totalizing explanations, privileged discourse, and boring encounter.

I invite you to take pride in being an artist. You are part of a great historical tradition of aesthetic performance, with roots that go way further than Vienna. Before psychology, psychiatry, and social work, there were performers whose creative expression healed others. They are the ancestors who deserve our remembrance and honoring. Today there are unknown performances on hidden stages spread throughout the land. Let us ask them to come forth and inspire the field to awaken from its comatose sleep. Nothing less is required for the creative resurrection of therapy as a living transformative art.

I invite you to make an apple nut for yourself—either in the way Christine Apple Nut did or in any other way. Perhaps you will feel inspired to draw a tiny apple on the corner of this page—don't forget the stem. Please do so right now and then cut it out, gently place it on top of your head, while wondering whether it will contribute to the nuts and bolts of your practice or make you a nut? Do not ask whether there is a difference.

CHAPTER 14

The Psychoanalyst Who
Wanted to Be a Priest

Part of my practice involves working with therapists who are stuck in a variety of ways. They may feel "burned out" by the profession, desire more creative life in their sessions, be looking for a way to evolve their own unique style of clinical work, hope for a boost of motivation or an injection of imagination, or want to get past a harbored dissatisfaction or hidden suffering. In the following case, we find how a psychoanalyst's pain and tears helped transform both therapist and client in a unique therapy of therapy. Suffering is no stranger to transformation—it can be one of our greatest teachers and contribute to evolving our creative edge.

This session was conducted in front of an audience of psychotherapists at a conference held in Brazil. I asked for a volunteer so I could demonstrate the "therapy of therapy." To my surprise, a distinguished older gentleman known as the "grandfather of psychotherapy" in Brazil started walking toward the stage. He was a psychiatrist and psychoanalyst who had mentored many therapists. Now we were going to have a conversation that aimed to help his clinical work be creatively transformed in a way that was meaningful to him.

Act I: "I, the Famous Psychiatrist and Psychoanalyst, Am a Failure"

Brad Keeney (BK): I am very moved that you chose to come up here.
Psychoanalyst: You are the person who is moving me greatly, deeply, and
        significantly. This morning, when I was in a Catholic church at
        the Santa Rosa de Lima Parish, I had a vision of your image.

My goodness, it was impressive! I saw your image and heard you say, "I bring you new parables." In this vision, you told me that an angel from heaven came with four bottles—a bottle of faith, a bottle of hope, a bottle of love, and a bottle of goodness. The angel said that all the bottles were free—that I could take what I wanted. Then the angel wrapped up the four bottles and handed me a tiny package; everything was placed in the palm of my hand.

BK: How interesting.

Psychoanalyst: Now I want to tell you about my professional career. As a psychoanalyst, I have found myself with many psychotics. I confess to you that I do not know how to work with them. I told one patient that I thought he was suicidal. He immediately spoke frankly to me, "I was going to kill myself and leave a note that you ordered me to kill myself." I was appalled. Soon after that case, a married couple saw me and, upon leaving, the wife said, "I am going to kill myself because you talked about suicide. I am going to kill my husband and my baby, and I want to leave behind a note in writing that Dr. S. orders people to kill themselves."

One patient after another has shaken me. Another paranoid schizophrenic once told me, "You are a lowlife, an indecent man, and dishonest. You ought to tear up your psychiatrist's diploma. Today you raped me in the street." I knew she was planning to call the authorities and, as she ran out of my office, I thought of calling my lawyer to send him to the police station to defend me.

Another patient came into my office with a revolver on his belt, sat down, and said, "I came here today to kill you." He insisted he was going to kill me and I thought I was experiencing my final hour. Then he stood up and asked me, "Are you crazy?"

I am an expert at giving psychoanalytic interpretations of all these crazy encounters, but I am telling you all of this because it helps me make another confession I haven't told anyone before. I have a paranoid schizophrenic son who is 30 years old, and that's the reason I became a psychiatrist. He believes he has run for the office of mayor in every major city in Brazil from São Paolo to Rio de Janeiro. Nothing works with him. I have tried every treatment that is known. I have great disappointment in my life. I have practiced different forms of therapy for 23 years and I feel I am impotent to get my son out of his psychosis.

Act II: Great Teachers

BK: I am struck by how remarkable it is that your life has brought you such powerful teachers. I know that you are a man who believes in God and appreciates the fact that God brings us our teachers. I heard every one of those patients you described as a distinguished teacher. In many ways they teach you. These were your teachers—not the books, not the university. Instead, these crazy people taught you. Maybe they were angels who came to teach you powerful things. The most powerful thing to learn is that every human being, and this includes you and me, is a murderer, a rapist, a thief, a fraud, capable of committing every horrible sin. We are all the same.

What I have been taught by the great healers from all around the world is that they feel that they are the same as everyone. I don't mean simple equality. I mean that they feel like everyone. They feel like a rapist. They feel like a murderer.

These people you told us about came to you and told you what you are. And you passed the test because you did not resist the truth. [At this moment, the psychoanalyst begins weeping.] In some way, you let this truth destroy you. By this I mean destroy some of your understanding, destroy some of your identity as a psychoanalyst. That was a lot of suffering. This is the true teacher.

I can't understand why God chose you to receive such powerful teaching, even to the extent that there has been a sacrifice of your own son. God sacrificed his son and theologically this has been a part of your life as well. I don't know why so much was given to you. I do know that the biggest teachings come with great pain and suffering.

In this response I utilized practically every metaphor the psychoanalyst had presented—teaching parables, angels with holy gifts, suffering with patients and family, and death. The fact that he began the session with a report about having been to church suggested he was a deeply religious man. I gave him the same report, or more accurately, a mirror-image reversal, where the patients were now angels delivering truths molded in the ovens of suffering.

Psychoanalyst: This is true.

BK: I see your body trying to have some kind of birth. It looks like birth when your pain comes up and tries to let something go. It tries to let something come forth. What are you trying to bring into the world?

The position and movement of his body and the expression on his face brought an image of birth to me so I shared that free association, with the consideration that perhaps it would help move us to another scene.

Psychoanalyst: In my heart, I am striving to do something good.
BK: Is it possible that you've become a Job of our time?
Psychoanalyst: I have no such pretension.
BK: You're being tested. You still have your faith and you still remain good. You came up here so you still have hope. And your gentle tears show us that love is still present. Your four bottles are free to everyone. You give them freely to your patients.

(The psychoanalyst physically shakes with a tremor of energy.)

BK: What is that? I see you. Is something trying to speak or come from your body? What is this?

Again I recycle his metaphors, but as I do so, his body starts shaking. This again reminds me of birth. I underscore this one more time.

Psychoanalyst: The pretension, the pretension to be powerful when I feel I am impotent in many practical instances.
BK: Is this power that doesn't realize itself?
Psychoanalyst: It is the power that is not aware of itself.
BK: So there's all this energy that comes and then stops.
Psychoanalyst: Exactly. I work with extremely difficult cases, including psychotics who are hard addicts. I feel that they do not accept my invitations. I feel I am impotent and that I cannot save one life that is lost, but I am obliged to accept this failure.
BK: So your life has brought you to what other people would consider the most impossible, hopeless situations, and yet it has not touched your power. Inside of you is all this power, but it doesn't come forth in a way that you think it should. Maybe you're not so sure. Maybe that is your power. Maybe this is the power that enables you to stand so firm where others would perish.

His power paradoxically enables him to survive situations where he is powerless to make a difference. This notion moves us from a hopeless situation where there is no power to one where it is a matter of how he brings his strength forth, that is, gives birth to its presence in a different way.

Psychoanalyst: Yes, this happens to me.
BK: What if there could be a moment in your life when you feel that all of this pain and all of this teaching had brought forth a break-

through. What would that moment be? What comes to your imagination?

Psychoanalyst: I would keep a suicide from killing himself.

Act III: The Psychoanalyst Who Wanted to Be a Priest

BK: Did you ever wrestle with wanting to be a priest?

This question simply dropped into my mind and I shared it. Though it is related to religious themes he brought up, in my imagination, I actually saw him wanting to be a priest as a young man. I would say that the "mind of therapy"—the pattern of interaction relating us—brought forth this idea and image.

Psychoanalyst: I thought about it when I was young. A friend of mine from the state of Maranhao said, "If you had a vocation, you would be going to a seminary. Your place is here in the world, among the people."

BK: I know you're a man with whom I must be completely honest. I will speak what I feel is the truth about you: I experience you as a priest who is hiding as a psychoanalyst.

This point is the fulcrum of the session. We have entered a different contextual home for holding the experiences of his life.

Psychoanalyst: I feel that there is a great truth to that. I feel that my many years of training were in order for me to be a lay father.

BK: When God sent you these people, I think they said to you, in their own unique way, "If you believe you're a psychoanalyst, then you are a murderer. If you believe you're a psychoanalyst, you're a rapist. If you believe you're a psychoanalyst, then this is death, suicides, and more death everywhere." But I think, without using words, through communication that speaks without saying, you tell them, "I am not a psychoanalyst. I'm here for a bigger reason. I'm with you for a bigger reason." Somehow they know this. But it hasn't reached your head yet. Your mind doesn't yet know it. Your body knows it. All the powers down there (BK points to psychoanalyst's chest) know it and, when your body trembles like it just did, your energy stops right there, at your head.

I am weaving together everything that has been spoken and presented in the session, doing so to bring forth a priest whose presence is a different order of message than the words uttered by a psychoanalyst. In other words, he has the presence, demeanor, and tone of a priest, with the talk of a psychoanalyst.

Psychoanalyst: I feel like you are explaining what I am seeing now, which was implicit. You are making clear to me what was obscure. These words of yours are words of profound truth. I fully accept them.

BK: You see, I think your faith is so pure that you are too humble in your mind to think that you could be a priest. Your problem is your humility.

Psychoanalyst: I feel that I cannot forego humility. I cannot forego humility.

BK: I don't think that you should because it makes you a better priest. You should continue being a priest without knowing what your truth is. Or you could let the energy come into your head and let you realize that you're one of the purest priests. However, this would take away your humility. If you'd realize that the energy you hold is the energy of a priest, I think you might be tempted to lose your humility.

This aims to help free him from the paradox that the consequence of allowing his energy to rise and come through his whole being, including his head, does not necessarily invalidate any mindful awareness of being an authentic priest.

Psychoanalyst: I feel a lot of confidence because I have an image in my mind of my paternal grandfather, who never left my eyes or heart. Last century (1800s), when he was 12 years old, my great-grandfather authorized him to go to French West Africa. He told me many wonderful stories. As a boy of 12, he was in the middle of lions and leopards. That man is in my blood with his counsel, and when the darkest and most difficult things appear and I feel I am going to be crushed, I feel his strength. He helps me feel that I won't be crushed because of the strength of his love.

Here, the psychoanalyst reveals what has been one of the major resources in his life: his grandfather. As he says, "that man is in my blood" and he "feels his strength" in the "darkest and most difficult" times. What the psychoanalyst did not know is that I had experienced a dream that very morning where I had seen my own grandfather and been awash in his blood.

Act IV: The Psychoanalyst Is a Secret Priest

BK: I understand this. The most important person in my life was my paternal grandfather. He was a man that everyone loved because he was a beautiful pastor of a church, and after he died, I began to feel things in a new way. Because he comes to me in dreams, I feel him inside of me. He always wanted me to be a minister, but I could never accept such a thing. My life, too, has been filled with much pain. Sometimes I didn't know whether I wanted to continue the struggle.

Early this morning, I had a dream of my grandfather. I didn't understand it. I'd like your help. Usually when I dream of my grandfather, he touches me with such love that I wake up weeping. This morning's dream was disturbing. Please know that he actually died of a cerebral stroke while bleeding from his mouth. This morning, in my dream, I was lying in bed with him. When I was a little boy, I visited my grandparents and I would sleep with him in his bed. Now, in my dream, I was sleeping with him as an adult and I was looking at him. How proud I was of him. He was so wise with words, had so much humor, so much love, and then all of a sudden, he just came near my neck and all this blood came out of him, spilling all over me. I was shocked. I am still shocked. I don't have a clue what this means. All morning I have been disoriented. At this moment, I present it to you for your help.

I turn the tables, asking the psychoanalyst for his help. I did not do so as a therapeutic maneuver. I was shocked by what he had said in relationship to my recent dream. I felt naturally drawn to ask for his assistance.

Psychoanalyst: What do I know? I could tell you some foolishness. The Chinese say that people cannot flee pain because pain gets you by the foot. To face pain is to triumph over it. People cannot escape the blood that came from the mouth of their beloved grandfather. I would say this: I want to receive all of this blood in me in order to overcome the blood that affected me so deeply in my life and from the man whom I loved the most.

While comforting me, he provides an answer to the question he is asking for himself: "To face pain is to triumph over it." What he tells me is therapeutic for me, but it is therapeutic to him as well. Notice how he articulates his advice in the first person voice: "I would say this: 'I want to receive … this blood in me in order to overcome the blood that affected me so deeply in my life and from the man whom I loved the most.'" He is unconsciously speaking of his paternal grandfather and son, though he consciously thinks he is addressing my paternal grandfather. He is also unknowingly speaking of my own son. I speak for both of us so he can answer for both of us.

BK: I still want to escape from the pain, but it still comes.

Psychoanalyst: I was thinking that you are a wonderful fruit of the pain of this blood that came out of your grandfather's mouth. The blood is life, and you are passing on much life to people. You are bathed in the life of the blood and, since the blood is your life, you will give more life to all of us.

BK: That touches me very deeply. Thank you for helping me. I always feel inspired by your presence in a room whether you talk or don't talk. You are a true priest for me. Thank you! [With tears, we both stand up and hug one another.] Yes, you are a true priest—to me and to others. Thanks for the teaching you have brought to us today. I am grateful.

At this moment, we were both deeply touched and feeling tears of suffering and joy. I felt that when he first talked, he was a psychoanalyst. When he didn't talk, he had the presence of a priest. Later in the session, he more deeply integrated these dual roles and moved toward more fully becoming a psychoanalyst who is secretly a priest. Now his words and silence both minister and heal.

In this example, the circular mind of therapy reversed the role of therapist and client so that the client, a psychoanalyst, became my therapist so he could be authenticated as a priest. At the time, I had been separated from my own son and had been trying to address numerous conflicted situations in my personal life. We each were a mirror for the other and, in our interaction, transformation was brought forth by the experiences we each shared.

The psychoanalyst now could be a priest without giving up psychiatry. He was able to more easily own the resources he held as a priestly presence in order to be a psychiatrist who administers hope, love, goodness, and faith to others. Whereas being a public priest would threaten his need for spiritual humility, being a secret priest humbly empowered his therapeutic practice. His life ministry of suffering could be utilized as a teaching to help him stand firm as a psychoanalyst in the face of the impossible circumstances of others.

CHAPTER **15**

Funny Medicines for Children

Humor and absurdity can also serve transformation. When my son was 12 years old, he started having nightmares about "scary bears." To help Scott overcome his fear, I asked him whether he thought other children might be afraid of bears.

"Yes, they must be," Scott replied. "And some of them are probably more scared of them than me."

"I bet you're right about that," I answered. "Maybe you should do something that would help them. Why don't you make them a medicine?"

"Yech, medicine tastes bad," Scott retorted.

"What about a 'funny medicine'? Something that would tickle them and get rid of their fears and worries."

"I never heard of a 'funny medicine.'"

"Me either. Let's invent one. Think you could do that?"

"Maybe."

Being a jazz pianist in addition to a university professor, I took my son to my recording studio to create a funny medicine CD for children who were scared of bears. Scott learned to use recording equipment he had previously not been allowed to touch, which thrilled him. Together, we wrote a funny, transformative story about a made-up bear. As Scott read the funny medicine out loud, I played an improvised musical background. That audio recording is included on the last track of the enclosed DVD. Here's the funny medicine we made for children scared of bears:

> This funny medicine is for all you children who are scared of bears. Get the largest piece of paper that you can find. Maybe you can find a piece of paper that is as big as your bed. You can even tape small pieces of paper together to make a larger piece.

On this large paper, draw a picture of the largest bear you can draw. Make its face look very scary and make sure you can see its teeth.

When you have drawn this bear, take a red pencil or crayon and draw a heart on its chest. Make sure the bear's heart is large enough to hold a picture of you. Draw a picture of yourself inside the bear's heart. This will make the bear become *your* bear. Your bear will have you in its life to protect you.

Now, take a piece of cloth or paper that is your favorite color and cover the bear's chest as if it were a shirt. Put the cloth on the bear, making sure that no one can see the bear's heart.

Take this bear and place it underneath your bed. When you go to sleep, know that your bear is protecting you. If any mean bear or monster wants to mess with you, it will have to go past *your* bear. And *no* monster can get past *your* bear!

Think of your bear whenever you are scared of anything. No monster or spooky thing can get past your bear. Make sure you keep your bear's heart a secret. Tell no one about your bear's heart and how your bear protects you.

After making that medicine, we kept on making funny medicines—medicines for children who were afraid of ghosts, worried about the dark, afraid of being left alone, concerned about the future, addicted to getting toys, nervous about how they look, troubled by their parents' future, among others. Here, the beginning acts were children's problems that were moved onto the middle performance stage that, in turn, carried us into a final act where the problem could be an opportunity to play with something once scary or troublesome that is now experienced as funny, delightful, and empowering.

We decided to submit our work for publication in a professional journal. After all, I was a professor at a university. We wondered where in the world would such absurd riddles be understood. We sent our creation to Japan, home of the Zen koan, and it was subsequently published in the *Japanese Journal of Family Psychology* (Keeney & Keeney, 1993). Scott, now known as "DJ Skee," became an internationally famous producer of hip-hop recordings in Hollywood, while I wrote a book entitled *The Creative Therapist*, which advocates reframing therapy as a performing art.

CHAPTER **16**

Creative Action Koans for Therapists

In this chapter, you will find absurd directives that help stop the "therapy habits" that interfere with and block creative movement in a session. These tasks "seed the unconscious soil" so as to facilitate the growth of transformative presence. Consider these to be incubation protocols that facilitate giving birth to inner creative resources. From this moment on, regard every session as a creative and experimental performance theatre for evolving both the creative edge and the imaginative transformation of all involved.

We can benefit from drawing upon absurdity to help us access creativity in a playful way. Performed absurd actions can tease, gently challenge, and effectively confound our overly rational professional minds, which desperately need to lighten up in order for more creativity to flow. I call these absurd prescriptions of action "creative action koans."

When facing an absurd (and rationally impossible) Japanese koan riddle such as: "What is the sound of one hand clapping?" you aren't supposed to think or say, "That is ridiculous; one hand cannot possibly clap." You are required to accept the legitimacy of this absurd and irrational invitation. Faithfully pursuing its answer may trip and free you from the constraints of habituated mind.

Think of "creative action koans" as absurd prescriptions to perform something that may seem as paradoxical, ridiculous, or illogical as a Zen riddle. Both aim to stop whatever is in the way of your being fully alive in each session. Take a deep mental breath, stop your self-evaluative chitchat, and jump in!

"Stop 'P-ing' on Me"

Now more than ever, we need to bring a new sense of imagination to therapy. I propose that we do this by going wildly absurd and absurdly wild. Let us, as therapy activists and dreamers, rediscover the universal archetype of the sacred fool who is capable of juggling realities and transforming fantasy into something unexpected and vital. This Coyote spirit can help guide us in many ways: by mixing up our rigid assumptions, instilling in us the hope of an underdog, or simply making us laugh when we most need it. The highest and most noble therapy is about tripping ourselves into seeing, hearing, and feeling the world with a different awareness. It should seek to offer an opportunity to have accidents of transformation.

Holy fools and jesters through the ages have always known that the first step toward liberation and enlightenment is to escape from lives that are overgoverned by the ideals of efficiency, predictability, control, and rationality. The essential ingredients of being creatively human are often upside-down, mirror-imaged, and reversals of common sense.

Crazy, absurd wisdom helps us question leaders who lazily invoke metaphors of scientism or patriotism or toss about arbitrary rules of maintaining a boring status quo masked as so-called "professional ethics." I invite you to challenge and overthrow the overimportance of the inflated metaphors of science, outcome studies, professionalism, and ethics that mask an enslaved obedience to mediocrity, or what Abraham Maslow called the "pathology of normality" (Maslow, 1971).

While delivering the opening keynote address to the American Counseling Association (in Honolulu, on March 28, 2008), I gave the audience of 3,500 counselors a task, the same prescription I will give you now. Right this very moment, grab a pen and write the letter "P" on your palm. Now draw an "X" over it to indicate the message: "Stop P-ing on me." When you hear a "P-word" trying to claim inappropriate authority, then hold up this hand so your palm can say, "Stop P-ing on me." The "P" word could be psychiatry, pathology, pharmacology, probability, professionalism, politically correct, postmodernism, psychobabble, piety, $p < 0.05$, and so forth. I urge you to do this with the intention of waking up your clinical practice. Start the revolution both within and outside yourself. Crazy therapy wisdom lets you tune in to the sounds of unknown prophets who dare us to love our contradictions, take care of our unknowing, dream with our feet, and dance wildly with our palms.

Join the Trickster Revolution

Most ancient cultures around the world value the foolish, but wise, tricksters—those who flirt during a solemn ceremony, laugh at a funeral, or weep at a joke. As the ancient Chinese sages might have advised the therapy profession:

- To become a whole therapist, twist your presence in clinical work.
- To become lined up with the creative flow, let your nature be bent.
- To become full of life in a session, empty yourself of all therapy ideas, assumptions, theories, and models.

With this wisdom in mind, consider organizing a campaign to annually appoint and support a trickster for every therapy conference. For the American Counseling Association, the anointed trickster might begin each conference by asking you to create a small card that has your name on it, followed by the letters ACA, which could be spelled out on the card as "Absurd Counselors Association" or as "American Contraries Assembly" or the never-ending "Adventures of Counseling's Awakening." For AAMFT, consider "Absent Absurdity Means Failed Therapy." And for either APA, think "Another 'P' Again."

I invite you to consider the patriotic playfulness of America's founding gadfly, Benjamin Franklin, who dared to tap lightning in a reckless experiment using his famous key and kite. What would happen if you tried to channel a similar wild, unpredictable bolt of energy into your clinical sessions and into every conference? Imagine what some creative transformative lightning could do for us. The resulting thunderclap might get our therapy heart beating again.

"Trickster Therapists of the World, Unite!" Let the revolution be sounded with nonsense, noise, uncertainty, and a wild fervor that invite a risen heart to playfully and creatively overthrow the hierarchies and exaggerated importance of static minds. I dare you to become fully human and fully alive. Now forget about whether you understand anything and instead take a stand and be as ridiculous and absurd and joyous and loving and creative as you can, doing so as if your life and clients were totally dependent upon it.

Drill a Hole in This Book

Consider this question: What are other ways could you relate to the book you are now reading? I invite you to drill a hole through this book. Yes, drill a hole in it. You may begin by pricking page one with the tip of your pen or pencil. Say out loud, "I have begun a new relationship with words, ideas, understandings, and theories." Consider that the hole in your book allows your book, as well as your mind, to breathe easier.

After you drill a hole through your book, ponder over what you have done. Then ask whether you are lost, wondering where you will find your answer to the challenges of therapy. Write out a simple request with a crayon, "Please help me find out how to make therapy come alive." Roll it up and push your question through the book hole before you go to sleep. Believe that your unconscious will deliver an answer. Your only uncertainty should concern the form in which the answer will be delivered. Will it come in a remembered or unremembered dream?

What's so different about this drilling? Haven't you been BORING for knowledge all these years? How many times have you seen the act of reading or studying to be a mission to extract knowledge from the text? Top Secret: There is no transport of knowledge. Knowledge is constructed rather than received. There is no data, only "capta." If you don't act, there will be no knowing.

Perhaps that hole is a blowhole. Take the book to a clinical session and squirt some water out of it. That would make a whale of a story for your clients and colleagues.

Four Fantasy Dreams for the Therapy of Therapists

I sometimes engage in what I call "fantasy dreaming" when I work with psychotherapists. Utilizing the metaphors they use to characterize their practice, clinical beliefs, and orientation, I am inspired to free associate the creation of a fantasy mini-dream episode. These fantasy dreams infuse absurd play into a woven multistoried context made up of a hybrid of nonsense and meaningful sense, all rooted to the client's cherished clinical premises. Inside this imaginary rendering, a directive is brought forth, evoked by the very creative process that created the context that holds it. What follows are four fantasy dreams for working with therapists and their ways of framing their clinical worlds. Feel free to modify them in any way that tickles your irrational curiosities and ripples your profound uncertainties.

William Tell and Client-Centered Therapy

I dreamt I was a client-centered therapist and an archer came along and asked why I hadn't placed a bull's-eye on every client I see. Dressed like William Tell, he asked, "Would you place the bull's-eye over their heart, belly, head, mouth, ears, hands, feet, or rear-end? How can you hit their center if you don't have a target to aim toward?"

Before I could tell him he was mad as a hatter, he picked me up as if I were an arrow, placed me in his bow, and shot me across the room. "Help!" I screamed as I flew right out the window of my clinical office. "Please help! Can anyone hear me?" I cried out as I zoomed across the landscape, across time, across all recognizable ways of being in the world I previously thought I knew.

"Whack!" I suddenly and authoritatively landed somewhere. I assumed I had hit the bull's-eye of a target I could not see because my head was stuck inside whatever I had pierced.

"Now you must discover the part of you that was the target. I shot you, but where you landed was determined by what center you called forth for yourself. Whenever you find your center, know that this is the center you must aim for with others."

"But what does this mean?"

"It means nothing until you find your bull's-eye," William Tell replied. "I will leave you with this advice: make a bull's-eye, a target. You know—circles around circles—on a small piece of paper and cut it out. Hang it on your office wall. Look at it at least three times for every session you host. Know that each time you look at this shown target, you will be shot again by Tell. Consider this the "show and tell" of creative therapy.

The Postmodern Bed

I dreamed that I was a postmodern therapist and met a cockroach at the end of time. Dressed with a tiny red cape, he was able to speak with a voice that sounded like Mighty Mouse, the cartoon character. "Here I come to save the day," he sang, as I approached him.

Though dizzy and confused by the whole scene, I managed to ask, "What is there to save? We are at the end of time."

"Exactly! Here I come to save the day," the cockroach sang again, but this time doing so while tap dancing on the most modern-looking clock I had ever seen.

As he continued singing and dancing, I realized that my mind was going into another dream world. I was dreaming that I was dreaming. I saw my bedroom with a huge bed in the center of the room. It had four bedposts that were so tall that they went straight through the ceiling, all the way through the roof and higher and higher into the sky and then past the sky, all the way to the edge of the universe. My first thought was that the bedposts were telephone lines going to the gods. As I formed a second thought, the red-caped, singing, and tap-dancing cockroach spoke again:

"These are the four posts of the four-cornered postmodern therapist. They go nowhere, which is the end of the edge. Never forget what I am telling you. In 30 seconds, you will wake up in a third dream world, but will believe you are awake in the same place where you first fell asleep. Know that you must immediately get up and make a plan to go to the store and purchase four coasters to place under each bedpost. On the surface of each coaster, cover part of it blue, so your coast has a sea. On the coast of your coaster, draw the outline of a horse. Now place these underneath each bedpost."

"Step back and look at what you have done. You, a four-cornered post-modern therapist, are now able to sleep and ride with the Four Horsemen of Apocalypso. Yes, that's right, I said 'Apocalypso.' Calypso, in case you have forgotten your Greek cultural ancestry, was a nymph and daughter of Atlas. She lived on the island of Ogygia, where once upon a mythological time, she discovered Odysseus washed upon her shore. She fell in love with him and promised she would give him immortality if he slept with her. He refused and was imprisoned for seven years until Zeus sent Hermes to set him free."

"Good God!" I shouted out, still asleep in a dream within a dream. "This is the craziest dream I've ever had. It must have been that trickster Hermes." I was awake in my dream because I was aware that I was dreaming but could not get away from the mighty little cockroach.

"Here I come to save the day!" It all started again. For the first time, I noticed that there was a full orchestra accompanying the bug whenever he sang. "Now you know and will never be able to forget that a four-posted bed can take you anywhere you want to go. But it will cost you seven years of your life as you try to figure out how to escape the temptation of being held by the promise of an immortal theory that claims to have the edge."

To this day, I am not sure that I have ever fully awakened from the haunt of that bizarre, multilayered, polyphonic, complexly voiced, diverse dream of dreams.

Jung at Heart

This third dream still has my Jungian colleagues confused. In it, I was floating across Switzerland and one famous therapist after another was flying near me. They each had their recognizable head, but the body of a bird. Freud flew by, as did Fromm, Fromm-Reichmann, Horney, and Reich. When they passed by my side, I could hear some music, what sounded like the song, "Young at Heart."

I must have seen at least several dozen therapy-birds fly by and sing that tune. Then a hot-air balloon began to ascend. As it came closer to me, the same song became louder and louder until POP, the balloon burst. Instead of falling, the balloon continued ascending. Or maybe I was dropping. Maybe I had been a balloon and I had popped. I was uncertain and this heavily disturbed me. It was then that the balloon passed right through me, with that song going on and on:

> Fairy tales can come true, it can happen to you
> If you're young at heart
> For it's hard, you will find, to be narrow of mind
> If you're young at heart
> You can go to extremes with impossible schemes
> You can laugh when your dreams fall apart at the seams
> And life gets more exciting with each passing day
> And love is either in your heart or on its way

As I started to laugh at the sheer absurdity of the situation that had arisen, I saw a cloud shaped as the face of Carl Jung. He wasn't laughing and he did not have his pipe. He appeared rather sad and without knowing why, I felt deeply sorry for him.

"Waaaaaa! Waaaaa!" This enormous sound blasted forth like a ship coming in from sea or a locomotive speeding along the rails. It made me jump, even though I was still falling in the sky. Without warning, a heavenly roaring voice proclaimed the sound of authority equal to any kingly lion that ever walked the African plains: "He, that man who is now a head in the clouds, did not build an ark. He misunderstood my instructions. Ark, not archetype! He shall forever be banished in his clouded head."

The voice went on: "You, young whippersnapper at heart, must never forget this song. Forget the archetypes and get on board the song. Learn it and then sing it, or play it or listen to it at the beginning of every clinical workday. Consider it the anthem of creative therapy."

Was It a Milton Dream or an Erickson Dream?

Up that cactus-scattered mountain I went, told to find a purple tomato. I thought I was in a dream because I was just on the edge of falling asleep when the scene began. I heard a soft voice that sometimes said it was Milton and at other times said it was Erickson. Did I hear it one way when I was awake and then another way when I fell asleep? Though I never met the fine doctor, I assumed that hearing the names of Milton and Erickson was important particularly since they seem to coexist in the different realms of dreams. At least that is what I was told by them in the dream.

This was the beginning of a daydream that came to me one sleepy, hot afternoon in Tucson, Arizona. Though it began as an afternoon reverie, I eventually did fall asleep and, in an authentic bona fide dream, found "the one" I was looking for all these years. In my real dream, I met a talking cactus. It was unable to walk, but full of barbs and piercing points. I don't believe I ever knew what a trance was until I had this dream. Perhaps it's better to say that I learned about being awake in a different way, so that by some twisted sort of logic—the cactus assured me—this meant I also knew about trance.

One of the saguaro's arms was holding a purple tomato. It too spoke, but not in any recognizable sound or language. It made unusual tapping sounds, sometimes like a clock and other times like a syncopated snare drum. When it tapped, I felt as if I heard words. That's what I mean when I say it talked.

"Tap, tap, tap, tap," it said, while I heard: "Yes, that's right. It's the galloping sound of the Four Horsemen of the Apocalypso. They'll make you all kinds of promises to take you far away from the familiar."

There was a pause.

"Tap, tap, tap," which I heard as, "Look up at the clouds and see the risen mind without an ark. It never learned to navigate sea or ground."

Followed by "Tap, tap," effortlessly translated as, "Make sure you know where you are aiming. This you were previously told. Aim for the center that has been sung forth."

Finally, the sharp staccato of a singular "Tap" and the beginning of an experience I am unable to adequately convey. It began with that purple tomato turning into a hot-air balloon that shot into the sky like an arrow, piercing a cloud, with a song I previously heard in another dream being played at full volume. It was an extraordinary collage of image, sound, and feeling.

Then it all stopped and the cactus dropped every single needle. When I looked to the ground the soil had changed to a wooden floor and there I saw the feet of my beloved grandparents and family. The soft voice, said, "Come home."

"Come home, where the rainbow begins and ends. Come home, where the beginning maternal voice once told you to come home. Take aim and come home. Travel far, but come home. This is your ark. This is where you will always be young at heart. Come home."

The cactus needles were scattered on the floor of my grandparents' house and they suddenly came to life, standing and dancing across the floor. I don't recall how long they danced, but it was an amazing show and I doubt I will ever witness it again. The needles effortlessly and smoothly glided along the floor, arranging themselves into letters of the alphabet that spelled out a message. I had to bend over to read the message spread across the well-worn floor:

Get a box of toothpicks and consider them your cactus needles. Say out loud to yourself, "With this lumber, these mighty little sticks, these imaginative cactus needles, I will build myself an ark." Paint the needles purple and use them to spell out the words, "Come Home." Place this on a mat of your choice, use glue to hold them on, and frame it. Now hang the camouflaged ark, your personal unsinkable Trojan horse that holds the creative words of transformation, on the wall of your clinical office. On the back of the framed work of art, draw a tiny cockroach with a red cape, having a cartoon bubble coming out of its mouth. Insert these words, as tiny words: "Here I come to save the day!" Never ever show this backside to anyone, but be prepared to make an exception every once in a while. Know that making this sign is the sign you have been waiting for all your life. It, and only it, can confirm that you have come home.

Mantra for Creative Therapy

Write down this quote from John Cage and place it underneath the chair you sit on during your clinical sessions. Call it your mantra for creative therapy:

> "Theatre takes place all the time, wherever one is, and art simply facilitates persuading one that this is the case."

Create an alternative business card for yourself. Here are several possibilities for what could be printed on it:

Apple Nut Therapist
Uncaged (thanks John Cage) Therapist
Director, Theatre of Creative Transformation
Builder of an Ark

A Prescription for Snap, Crackle, and Pop!

I conducted a session with a middle-aged chronic worrier whose therapist could not get her to stop talking in therapy. In his previous hour-long meetings with her, he rarely was able to contribute more than a few minutes of talking time. In my time with them, I listened to her talk for about 5 to 10 minutes and then held up my hands and authoritatively said, "Stop!" I bent over, caught her attention, and immediately asked, "Let's add 5 to 10 minutes of something new to every one of your days; what would you like for that to be?"

She replied, "I'd like to feel some calm."

Pointing out how much work she gives to trying to figure things out with her mind, I said that she needed some way of distracting her inner talk, ceaseless analyzing, voiced concerns, ongoing discussions of the purpose of life, and seemingly infinite plans for action. As I put it, she needed an "oasis of calm" where her thinking, worrying, and sorting things out could be given a rest. At that moment, I got up from my chair, went to my desk, opened a drawer, and pulled out a pen and a small yellow pad of paper.

"I'm going to write you a prescription. Take this medicine three times a day and it will give you what you have requested." I wrote out a prescription, signed it, as did her therapist, and handed it to her:

> *Three times a day, pour some Rice Krispies into a bowl. Add milk. Then get your ears as close as you can to the bowl and focus all your attention to LISTENING to the "snap, crackle, and pop!"*

We discussed how she might get ready to take her medicine. Before she went to fetch her cereal, she could start singing a song. Music, I explained, would interrupt her worrying and get her nicely prepared for hearing the cereal sounds and rhythms. Because she was a religious person, we asked whether she wanted to say grace before she poured the milk. Without hesitation, she agreed that a prayer would be a good idea. We discussed how she would pray and what she would request. The conversation centered on her asking to be in a state of readiness, openness, and receptiveness to what she might experience and learn from the sounds of the cereal coming alive.

Rice Krispies subsequently became the teacher this woman had been seeking. The various preparations for getting her in a good position to listen were as important as the sounds that taught. When she was ready—her inner and outer bowls were empty—she could hold, hear, feel, see, learn, and feed off the random, chaotic sounds of noise. Preparation for the *snap, crackle, and pop* helped stop her world and be available for entry into a transformative realm of creative experience.

I invite you to write a prescription (to yourself) to take some Rice Krispies. As you move toward this task, pause and existentially consider how rice feeds so many people in the world—it literally has been the fuel for keeping most of humanity alive. Also contemplate how milk provides a bond with mothers who deeply care for their children's health and growth. In addition, meditate on how an empty bowl is required for a filling. Finally, reflect on how transformation is at the heart of all these matters.

Seriously, or lightheartedly, consider the creative ways you can prepare yourself to put some Rice Krispies into a bowl before a clinical session, how you can carefully and respectfully and absurdly pour some milk, and how you can get close to the surprising sounds that are ready to teach you how to add some *snap, crackle, and pop* to each of your sessions. Become the rice that explodes with creative surprise as you foster and mother the presence of creative expression in each and every session.

Snap goes the sound of each release—the openings, entries, and passages to more creative and resourceful abodes of experience! *Crackle* goes the back-and-forth, up-and-down, in-and-out motions of the breathing, walking, and dancing of transformative movement! *Pop* goes the fully embodied birth of the absurd theatre of the divine, where therapists and clients find their way to *bringing forth the moments when the gods play us!*

Giggle with the universe as you wonder whether the transformative therapy of your therapy is as close to you as a bowl of Rice Krispies. Consider condensing the entire message of this book to three sounds: the unexpected sounds and rhythms and movement of rice clapping its applause for your entry into creative therapy. Honor the way in which an empty bowl prepares you for the fullest empty teaching. To awaken your clinical sessions, make way for performing the irrational art of *Snap! Crackle! & Pop!*

CHAPTER **17**

The Therapeutic Crossroads

Though most outsiders have heard of New Orleans, fewer know about the more isolated and deeply impoverished communities near our Mississippi River Delta—towns such as Lake Providence, Tallulah, St. Joseph, Rayville, Goodwill, Delhi, Forest, Pioneer, and Waterproof. This is where the blues was born and where it is lived today. Here among the cotton fields, swamps, and juke joints is a wild frontier of creative therapy.

After my wife and I settled into this part of the country, I quickly tracked down a retired ethnographer named "Junior," who over the years had frequented the remaining juke joints while interviewing the old maestros who shouted out the blues. I found him living in a well-worn shack in the backwoods. His dilapidated blues-mobile sat rusting in the front yard. Though it no longer goes anywhere, Junior's memory and stories can take you to every juke joint in the Delta.

For example, "Disco 86" is a juke joint in Waterproof, Louisiana, where Annie Mae McKeal cooks fried catfish, fried chicken, fried pickles, and chitlins for people to eat while sitting under a big sweetgum tree. Here in Waterproof, three out of four African Americans live *below* the poverty level. Junior says that there's a rule for survival that must be remembered: "Never loan anybody your toothbrush, your shotgun, your wife (or husband), or your blues-playing equipment—not necessarily in that order." The blues legend, Little Milton, recorded a song about "Annie Mae's café." This is the place. It's in the hood where I work.

Junior gave me the names and addresses of the blues old-timers and told me how to be careful as I ventured into the hidden, remote places of the old Deep South. To my surprise, he whispered that he had been taken to the legendary and assumed mythological "crossroads." "Yes," he told me, "there is an actual place where musicians can go and get baptized by the spirit of the blues."

Some call it a "meeting place with the devil," but I think it is where you make your soul available to voice the blues, whether they be strummed on a guitar or spoken in a clinical session. It's where you come to make your-self available to help "call life forth."

Junior was taken to this sacred site by some elder bluesmen, and there he found a small stone in the very center of the crossing. He gave me that stone as a welcoming gift. With it, he christened the beginning of my jour-ney into the heart of southern mojo, gumbo cures, and swamp wisdom.

Soul and creativity flourish in the southern swamps, and in Louisiana, there is something extra that we call *lagniappe*. It includes an invisible gift—something felt that gives life an extra kick. It is no surprise that jazz, blues, zydeco, and gospel came out of this wilderness of the soul. Here my friends and I have commenced a quest to bring forth creative ways that help people transform their lives. In the crumbling shanties, inside the barbed-wire prison walls, and in remote and relatively unheard of social service and university sliding-scale clinics, we do our work. Here we have learned about the transformative ways people can overcome impoverished circumstances and suffering.

Therapists, counselors, social workers, and other dedicated people help-ers need to move past the limitations of their professional models and the-ories. Our work needs to shine light on a new highway—one that leads us away from therapy but helps us become more therapeutic. Like Louisiana's food, music, and celebrative expression, the way forward is through the creative spirit, fired by a passionate heart and syncopated soul.

Welcome to the road that takes you past all therapies. If you'd like, you can say that it will carry you to "the gumbo cure." You make gumbo by uti-lizing what you already have in the kitchen and then blend it all together, while adding some spice and hot sauce to make it come alive. Consider doing the same with your clinical cases. Utilize what your clients bring to you. Add some spice through wild talk and playful imagination. Then throw in some hot sauce through the ways you initiate unexpected action.

We know the old ways down here. Furthermore, we are beginning to fully appreciate how the ancestral cultures knew that the only way to make you a creative therapist is to throw you into a big pot of transformation and then cook you. That's what the Bushman *n/om-kxaosi*, the oldest healing culture in the world, say in the African Kalahari: You have to get cooked to become a healer. It's also how we understand the deeper stirrings of gumbo: You have to get stirred, shaken, and cooked on a long, slow heat to become a creative therapist.

While getting cooked, you remember what the Guarani shamans of the Amazonian forest teach: The greatest temptation is not to be consumed by anger, even when you are justified for expressing endless rage. Becoming a "heckler" who interrupts speeches at conferences voices no wisdom. It is a "grave" mistake that creates a "wall" of oppositional division. We must surrender our hatred to the boiling pot.

In the Kalahari pot of the gods, we remember the teachings of the Bushmen mothers: Healing is always about opening our hearts to other opened hearts. Betty Alice Erickson proposed the same for what was essential about the transformative presence of Milton H. Erickson. His sincere invitation to make a connection is what held his sessions. As Betty Alice describes this: "It was Dad accessing his pure unalloyed love for people and his belief in them, setting this love and belief, so to speak, between them" (Erickson & Keeney, 2006, p. 44). This provided a channel for connection, an atmosphere for entering into communion and fullest presence.

Please consider the discourse about therapy presented in this text as an attempt to usher a therapy of our professional discourse, a feeble effort to heal the ways in which our cherished beliefs too often blind us and lead us to recklessly promote another fundamentalism with all the terror and havoc it advocates. I confess that my own words end up being another narration, another way of wrapping words around the subject, and it, too, must be loosened and released.

From the Kalahari Bushmen women healers, the *n/om-kxaosi*, I learned that words are always to be mistrusted. They are the play of tricksters and can be both helpful and a hindrance. They are helpful, even a medicine, when they are used to pry us free from the knots that interfere with our being moved by life rather than fixed ideas. Our words are true for the moment that brought them forth; then they must be released to the wind and blown away. We return, as we should over and over again, to the stage of action—spontaneously performed words unscripted by narrative.

Go ahead and stuff yourself with any ideas you like, from anthropology to sociology, psychology, phenomenology, and phrenology; from culturalism, feminism, narrativism, and postmodernism to postculturalism, postfeminism, postnarrativism, and post-*post*ivism—add every doggone -*ism* and -*ology* your heart desires to feast on. Throw it all in the big pot and maybe we'll add some alligator, crawfish, red beans, and rice, along with our secret Louisiana red-hot therapy sauce. After we get done with you, you'll be all cooked—a nice gumbo blend of swamp complexity, bayou contradictions, jazz cacophonies, soulful harmonies, blues lines, spirited tap dancing, and homeboy and homegirl therapeutic charm. Perhaps you'll help us bring forth the posttherapy future of the many theatres of creative transformation. We assure you that all of this will be deliciously creative and mighty tasty.

Louisiana is a wild state of mind that graciously and absurdly hosts gumbo cures and swamp wisdom. Jump inside our transformative pot and find yourself in the middle of the therapeutic crossroads. There, every-thing may be utilized and juxtaposed so as to create a motor for growth, a buzzing intersection of change, a progression through lived contraries. Come on down for a walk and dance on the therapeutic crossroads.

I prefer that the performance of therapy have the final word. I am writing this case story following a day working in the Delta towns near where I live. Today I saw a mother and her 16-year-old daughter, who had been court ordered for therapy due to the girl's outspoken threats to kill her mom and run away. Last week, I had gone to their trailer home to see the mother. My first reaction upon entering the trailer was to declare, "My, oh my, I have never seen such a neat place in all my life." Every room was in perfect order and totally free of dust—fit as a fiddle and clean as a whistle. My colleagues and I sat down with the mother and heard about the family conflict and then asked whether the main disagreement was over the daughter keeping her room clean. Mother, who makes a living as a house cleaner, said that was indeed the topic that always gets them going at each other's throats.

I asked the mother whether she would conduct a little experiment that day. I invited her to tilt a couple of picture fames and see if her daugh-ter noticed it. If the daughter said something, mother was simply to reply, "This is my way of honoring you." The mother replied that she had actu-ally thought about doing something like this to see if her daughter would notice anything that she does in the house. In a burst of enthusiasm, she sprung up from her seat, went to the dining room area, and turned a paint-ing of a zebra upside down. She then whipped around and went to the kitchen and turned a metal sculpture of a rooster upside down.

Following that meeting, we immediately went to the high school to see the daughter. After we made a connection, she also was invited to try an experiment that same night. She was to make sure that her hands and fin-gernails were absolutely immaculate (as they already were) and then dip one finger—any finger that she chose on her own—into some mud, mak-ing it dirty as possible. With nine clean fingers and one dirty one, she was to see if her mom noticed. If her mother said anything, the daughter was to reply with, "My other nine fingers are honoring you." She seemed quite excited about this and remarked that her best friend liked to put a colored "dot" on her face just to see if anyone at school would notice.

Now, a week later—the very day in which I am writing this account—I invited the mother to come to the school where my team would see both of them in the high school yard. To our surprise, it was immediately apparent that the two had become best buddies. Holding hands, they giggled as they reported how much laughter had come into their lives. They were espe-cially proud to discuss how they enjoyed watching the reactions of other visitors to their home—who noticed, who didn't notice, who was upset, who laughed, and so forth.

I encouraged them to continue this practice, setting aside a time each day to plan what they would alter in the house, or in their speech, dress, or meals, before each visitor came over. They were to collect stories about the ways their home was able to bring forth surprise and laughter. When it was time for class, the daughter announced that she wanted to write down these stories and create a best-selling southern novel. Her mother replied, "I think she could become a great writer." The daughter then asked her mother to come to class with her, but she had to get back to work. They embraced and giggled before going their own ways.

We can learn from these clients. Mess with the words, rearrange the meanings, tilt the frames, turn things upside down, create some disorder, and play with whatever takes place inside and outside our clinical sessions. Do so with the curiosity of a southern novelist or playwright. Let's see if we can bring some more surprise, laughter, life, and transformation to our clients, ourselves, and the profession.

Today I also visited, for the first time, some new clients who live in a fallen-apart shack that the state has examined four or five times to see whether it should be condemned. Everyone in the neighboring communities talks about the place because, believe me, it looks like a tornado blew through it. The front yard is full of junk, including an old dilapidated tractor that hasn't been used in decades, heaps of metal, and other odds and ends that would not be identifiable by Sherlock Holmes. The inside of their home is dim, with only one light, a mattress on the floor next to the kitchen, and six mangy dogs, with rats and cockroaches zooming around as you talk—you get the picture.

I am there, in part, because all the local authorities think there is something going on with this family. Their teeth are rotten, the parents don't work, and the children don't always get to school. When I went in, I expected to meet the Addams Family, wondering whether a hand would crawl across the floor. After we greeted one another, I said, "You know, you folks have what money can't buy most people. You have an audience. I bet you are the most talked about family in this neck of the woods."

"Yep," Mom proudly agreed.

"The problem I see here is that your life is too boring. Why don't you really give them something to talk about? You could raise that tractor up so it looks like a dog standing up, maybe even paint it to look like a big animal. Or you could make an elephant-sized statue out of all that metal." As I said this, I noticed a collection of bronze giraffe figures in the living room. They were barely perceptible because of the darkness. "Are those giraffes?"

"Yes, those are our giraffes. Let me show you how I can communicate with them," the oldest adult daughter replied, with a baby slung over her shoulder. She jumped up, got the giraffes, and placed them in an arrangement on the kitchen table as if demonstrating a form of play therapy.

Expressing amazement, because I was rather astonished at this unexpected display of a therapeutic practice, I replied, "See, you have been holding back. You all are creative. Imagine if you could place a giraffe in your front yard."

Mom said, "I have a unicorn collection."

"What?"

"I collect unicorns."

"Where?"

"They're in a box in the bedroom."

I smiled and began, "You've been holding out on everyone. All this creativity is boxed up and you have a stage in your front yard where people drive by wondering what you're up to. Those unicorns are from the world of imagination and you need to release them."

Mom interrupted, "Did you know that they believe the unicorn was a kind of goat?"

"No, I did not know that. And I did not know that you have a lot of knowledge about unicorns. Do the kids know about your unicorn expertise?"

"No, I studied about unicorns a long time ago."

This was the moment for transformation. Moved by the immediate situation, I proceeded to say, "I want to invite you to bring the unicorns out of the box. Tonight and tomorrow, I think you should consider placing a huge sign in your yard that says, 'UNICORN RANCH.'"

"That's funny. The kids did put up a sign about wrestling that got everyone to talking.'"

"Excellent. I think you should put up many signs, with some having arrows pointing to the unicorns."

"You bet we're gonna do that. That's awesome!"

"You're a family that needs to get more creative and more interesting. You know, if you put up enough signs and spread enough rumors, I bet one day that some folks will drive by your house and claim that they saw a unicorn. Before you know it, the whole town will say there are unicorns and people will start to come from far, far away, looking for unicorns."

"Yeah, just like Bigfoot."

"Yes. That's right. You see the unicorn is from a magic world. It comes to life when people believe in it, announce it, and start hunting for it."

Mom, with the family all smiling and giggling, replied, "When you come back next week, you're going to be surprised. This will be a unicorn ranch. I never thought of anything like that before. Where did you come from?"

"You see, you already have a stage, an audience, creativity, imagination, and a box of unicorns. Now all you have to do is set them free."

All therapists, like this crazy Louisiana swamp family, live in a word mess, piles of theories strewn here and there, with a tractor in the front yard that hasn't tilled any soil for a long while. But, somewhere inside of you is a box of creative treasure just waiting to be opened and placed on your stage. Yes, creative therapy is no more, and no less, than bringing out the unicorns.

References

Bakan, D. (1967). *On method: Toward a reconstruction of psychological investigation.* San Francisco: Jossey-Bass.

Barker, P. (2008, March 17). Ghost writer. *New Yorker,* 41.

Barron, F., Montuori, A., & Barron, A. (1997). *Creators on creating: Awakening and cultivating the imaginative mind.* New York: Jeremy P. Tarcher/Penguin.

Bateson, G. (1972). *Steps to an ecology of mind: Collected essays in anthropology, psychiatry, evolution, and epistemology.* New York: Ballantine.

Bateson, G. (1979). *Mind and nature: A necessary unity.* New York: E.P. Dutton.

Bateson, G., & Bateson, M. C. (1988). *Angels fear: Towards an epistemology of the sacred.* New York: Bantam.

Brand, S. (1974). *II cybernetic frontiers.* New York: Random House.

De Mille, A. (1991). *Martha: The life and work of Martha Graham.* New York: Random House.

Englar-Carlson, M. J. (2003). Enough about models and abstractions, let your therapeutic soul be free: An interview with Bradford Keeney, *Family Journal, 11*(3), 309–314.

Erickson, B. A., & Keeney, B. (Eds.). (2006). *Milton H. Erickson, M.D.: An American healer.* Sedona, AZ: Ringing Rocks Press.

Erickson, M. H. (1980). *The collected papers of Milton H. Erickson.* New York: Irvington.

Field, S. (1979). *Screenplay: The foundations of screenwriting.* New York: Dell.

Franchi, S., Güzeldere, G., & Minch, E. (1995). Interview with Heinz von Foerster. *Stanford Humanities Review, 4*(2).

Geuens, J.-P. (2000). *Film production theory.* Albany, NY: SUNY Press.

Glendinning, C. (1994). *My name is Chellis and I'm in recovery from western civilization.* Boston: Shambhala.

Goldwater, R. J. (Ed.). (1945). *Artists on art: From the XIV to the XX century.* New York: Pantheon.

Held, R., & Hein, A. (1963). Movement-produced stimulation in the development of visually guided behavior. *Journal of Comparative and Physiological Psychology, 56*, 872–876.

Herrigel, E. (1971). *Zen in the art of archery.* New York: Vintage.

Johnson, S. (2001). Family therapy saves the planet: Messianic tendencies in the family systems literature. *Journal of Marital and Family Therapy, 27*(1), 3–11.

Jung, C. G. (1973). *Answer to Job.* Princeton, NJ: Princeton University Press.

Keeney, B. (1982). Not pragmatics, not aesthetics. *Family Process, 21*(4), 429–434.

Keeney, B. (1983a). *Aesthetics of change.* New York: Guilford.

Keeney, B. (1983b). Size and shape: The story of a conference. *Journal of Strategic and Systemic Therapies, 2*(1), 72–79.

Keeney, B. P. (1990). Un metodo per organizzaire la converszione psicoterapia. *Terapia Familiare, 32*(March 1990), 25–39.

Keeney, B. (1990). *Improvisational therapy: A practical guide for creative clinical strategies.* New York: Guilford.

Keeney, B. (2005). *Bushman shaman: Awakening the spirit through ecstatic dance.* Rochester, VT: Destiny.

Keeney, B. (Ed.) with I. W. B. A. Mekel. (2004). *Balians: Traditional healers of Bali.* Philadelphia: Ringing Rocks.

Keeney, B., & Keeney, S. (1993). Funny medicines: Improvisational therapy with children. *Japanese Journal of Family Psychology, 7*(2), 125–132.

Keeney, B., & Ross, J. M. (1985). *Mind in therapy: Constructing systemic family therapies.* New York: Basic.

Keeney, B., Ross, J., & Silverstein, O. (1983). Mind in bodies: The treatment of a family that presented a migraine headache. *Family Systems Medicine, 1*(3), 61–77.

Kottler, J., & Carlson, J. (Eds.). (2005). *Their finest hour: Master therapists share their greatest success stories.* Boston: Allyn and Bacon.

Kottler, J., & Carlson, J. with B. Keeney. (2004). *American shaman: An odyssey of global healing traditions.* New York: Brunner-Routledge.

Kottler, J. A., & Carlson, J. (Eds.). (2003). *The Mummy at the Dining Room Table: Eminent therapists reveal their most unusual cases and what they teach us about human behavior.* San Francisco: Jossey-Bass.

Madanes, C. (1981). *Strategic Family Therapy,* San Francisco: Jossey-Bass.

Malone, T. P., Whitaker, C. A., Warkentin, J., & Felder, R. E. (1961). Rational and nonrational psychotherapy. *American Journal of Psychotherapy, 15*(2), 212–220.

Maranhao, T. (1984). Family therapy and anthropology. *Culture, Medicine, and Psychiatry, 8*(3), 255–279.

Maslow, Abraham (1971). *The Farther Reaches of Human Nature.* New York: Viking Press.

McCulloch, W. S. (1965). *Embodiments of mind.* Cambridge, MA: M.I.T. Press.

McGoldrick, M., Pearce, J. K., & Giordano, J. (1982). *Ethnicity and family therapy.* New York: Guilford.

Montuori, A. (2003). The complexity of improvisation and the improvisation of complexity: Social science, art and creativity. *Human Relations, 56*(2), 237–255.

Montuori, A. (2005). Gregory Bateson and the promise of transdisciplinarity. *Cybernetics and Human Knowing, 12*(1–2), 147–158.

Montuori, A., & Purser, R. E. (1997). Le dimensioni sociali della creatività. *Pluriverso, 1*(2), 78–88.

Nachmanovitch, S. (2005, February). On teaching improvisation. A talk with college and university conductors at the College Band Directors National Association, New York.

Neill, J. R., & Kniskern, D. P. (Eds.). (1982). *From psyche to system: The evolving therapy of Carl Whitaker.* New York: Guilford.

Osberg, D. (2006). *What's the three-act structure?* Available from settingthingswrite. com.

Papp, P. (September–October, 1984). The links between clinical and artistic creativity. *Family Therapy Networker, 8*(5): 20–28.

Papp, P. (1983). *The process of change.* New York: The Guilford Press.

Selvini Palazzoli, M., Cecchin, G., Prata, G., & Boscolo, L. (1979). *Paradox and counterparadox: A new model in the therapy of the family in schizophrenic transaction.* New York: J. Aaronson.

Ray, W. A., & Keeney, B. P. (1993). *Resource focused therapy.* London: Karnac Books.

Reuther, R. R. (1992). *Gaia & God: An ecofeminist theology of earth healing.* San Francisco: Harper.

Reuther, R. R. (n.d.). *Ecofeminism.* Retrieved Oct. 9, 2008, from www.spunk.org/ texts/pubs/openeye/sp000943.txt.

Spretnak, C. (1993). Critical and constructive contributions of ecofeminism. In P. Tucker & E. Groves (Eds.), *Worldviews and ecology: Religion, philosophy, and the environment* (pp. 181–189). Philadelphia: Bucknell.

Spencer-Brown, G. (1973). *Laws of form.* New York: Bantam.

Stenmark, L. (2001). An ecology of knowledge. Paper presented at the November 2000 Academy of Religion conference, Women and Religion section, Nashville, TN. Available from metanexus.net

Thompson, W. I. (1977). Meditation on the dark ages past and present. In N. J. Todd (Ed.), *The book of new alchemists* (pp. xx). New York: Dutton.

Turner, S. (2005). *Story, script, and the spaces in between.* Retrieved September 10, 2008, from www.studycollection.co.uk/storyscriptsspaces.html

Varela, F. J. (1992). *Ethical know-how: Action, wisdom, and cognition.* Stanford, CA: Stanford University Press.

Von Foerster, H. (1987). Cybernetics. In *Encyclopedia of artificial intelligence* (pp. 225–227). New York: Wiley-Interscience.

Von Foerster, H. (2003). *Understanding understanding: Essays on cybernetics and cognition.* New York: Springer.

Wátzlawick, P., Weakland, J. H., & Fisch, R. (1974). *Change: Principles of problem formation and problem resolution.* New York: Norton.

Wegman, R. (2008). *Eco-Sophie: Thinking and knowing in dynamic harmony with nature's ways.* Doctoral dissertation proposal, California Institute of Integral Studies.

Whitaker, C. A. (1975). Psychotherapy of the absurd: With a special emphasis on the psychotherapy of aggression. *Family Process, 14*(1), 1–16.

Whitaker, C. A. (1976). The hindrance of theory in clinical work. In P. J. Guerin (Ed.), *Family therapy: An introduction to theory and technique* (pp. 154–164). New York: Gardner Press.

White, J. E. (1997, August 15). The poorest place in America. *Time,* Volume 166, pp. 35–36.

Whitman, W. (1900). *Leaves of grass.* Philadelphia: David McKay.

Wolf, A., & Schwartz, E. K. (1958). Irrational psychotherapy: An appeal to unreason. I, II, & III. *American Journal of Psychotherapy, 12*(2–4), 300–314, 508–521, 744–757.

Woolf, V. (1926). The cinema. *ARTS, 9*(6), 314–316.